Southamerican
Cichlids I

Ulrich Glaser sen.
Wolfgang Glaser

Verlag: A. C. S. GmbH, Germany

Vorwort
Foreword

Unser erstes Buch
„LORICARIIDAE all L-NUMBERS" war ein Riesenerfolg und wir arbeiten weiter.

So ist auch dieses Reference-Heft nur ein Auszug aus unserem Gesamtwerk mit Bildern über alle Zierfische aus allen Erdteilen.

Unser Ziel ist es, übersichtliche Bildbände zu erstellen, die es jedem Aquarianer,
aber auch Händlern und Importeuren, auf einfachste Weise ermöglichen, *jeden* Fisch schnell und problemlos zu identifizieren und anhand des kurzen Symboltextes die wichtigsten Pflegegrundsätze zu erkennen.

Unsere Bücher enthalten in übersichtlicher Reihenfolge jeden bis jetzt bekannten Zierfisch nach einem genauen Code-Nummern-System geordnet, wobei jede Art ihre Code-Nummer hat und immer behält, selbst dann, wenn sich der Artname eventuell einmal ändern sollte.

Darüberhinaus wird jeder neu entdeckte oder neu gezüchtete Fisch jährlich in Bild und Kurztext als Ergänzung erscheinen, zusammen mit Infos über Namens-Änderungen etc., womit Sie mit unserem Werk dann immer "up to date" sind.

Dank unseres ausgeklügelten 7-stelligen Code-Nummern-Systems hat in Zukunft jeder Zierfisch an jedem Ort der Welt seine eigene Code-Nummer für immer, sodaß sich unsere Bücher auch hervorragend und international als Katalog und Preisliste für den Handel eignen.

Unser Werk soll keinesfalls allgemein- oder artbezogene Fachliteratur überflüssig machen, sondern wir wollen Ihnen nur die unvorstellbare Arten-Vielfalt übersichtlich präsentieren und Ihnen die Möglichkeit geben, schnell und problemlos Ihre Fische zu bestimmen, für Pflege, Zucht etc. nehmen Sie sich bitte gute Fachbücher zur Hand, von denen es genügend gibt.
Wir wünschen Ihnen viel Freude mit diesem Aqualog-Buch, Kritik immer willkommen, wir werden Ihre Anregungen in der Neuauflage berücksichtigen.

Our first book "LORICARIDAE, all the L-numbers" has been a huge success and we are now continuing the work.

This reference volume is also just one part of our comprehensive reference work which contains photographs of all the ornamental fish from every part of the globe.

**It is our goal to produce clearly laid out illustrated volumes making it possible for every aquarium owner, retailer and importer to identify any fish quickly and simply and to obtain basic care information about them using the short symbol-based texts.
Clearly arranged, our books contain every known ornamental fish, listed according to a precise code-number system in which each species is allotted a code-number, which remains the same, even if the name of the species changes.**

**Furthermore, each year a supplement will be published containing a picture and basic information on newly discovered or newly bred fish including information on name changes etc.
This means you will always be up to date with our reference work.**

**In the future, thanks to our ingenious 7-figure coding system, every ornamental fish in every part of the globe will have its own permanent code-number making our books eminently suitable as the basis for international catalgues and price-lists for the trade. This reference work is in no way intended to make other general or specific ornamental fish reference works redundant.
Instead our intention is to present the unbelievable variety of species in the clear format, enabling you to identify your fish quickly and simply. For information about care and breeding please consult any of the good specialist guide books, of which plenty
are available.**

We hope you enjoy using this Aqualog-Book. We are always pleased to receive your comments and criticism; this will then be taken into consideration in the new edition.

Mörfelden-Walldorf, Januar 1996

Ulrich Glaser sen.
Wolfgang Glaser

South American Cichlids I
Was sind Cichliden???

Cichliden und ihre „Verwandten" kommen in allen 5 Erdteilen vor, die meisten und farbenprächtigsten in Afrika und Südamerika. Unter Cichliden versteht man in Deutschland „Buntbarsche". Bei einem „Barsch" denkt man bei uns in Europa unwillkürlich an einen grauen Speisefisch, wobei man aber nicht vergessen sollte, daß es auch hierzulande sehr schöne Arten, wie zum Beispiel den Flußbarsch (Perca fluviatilis), oder die ursprünglich aus Nordamerika stammenden Sonnenbarsche (Lepomis) gibt, die es in Bezug auf Farbenpracht ohne weiteres mit ihren Verwandten in aller Welt aufnehmen können, aber auch dort sind die größer werdenden Arten ebenfalls begehrte Speisefische.

Da in den Aquarien meist die bunten Exoten gepflegt werden, spricht man (in der Deutschen Aquaristik) von „Buntbarschen".

Bei den Barschen gibt es eine unvorstellbare Artenvielfalt, es gibt „Winzlinge" wie Elassoma evergladei, die ausgewachsen gerade einmal 3cm erreichen, oder „Riesen", die bis zu 100cm groß werden.

Sie können lang gestreckt sein, wie z.B. die Hecht-Buntbarsche / Pike-Cichliden, oder ovale typische „Fisch-Form" haben, oder auch bizarr, oder rund-tellerförmig wie die Discus-Buntbarsche und Scalare.

Die Cichliden gehören allgemein zu den höher entwickelten Fischen, sind meist, aber nicht ausschließlich, revierbildend, es gibt vollkommen „friedliche", die sich ohne weiteres für ein gemischtes Gesellschaftsbecken eignen, mehr oder weniger „agressive", die keine andere Art neben sich dulden, oder die „streitsüchtigen", die schon untereinander echte Rabauken sind.

All das Vorgenannte trifft in erster Linie für die „Southamerican - Cichlids" zu, man sollte also unbedingt darauf achten, welche Arten man zusammen pflegen will, ob sie auch zueinander passen und harmonieren.

Nicht nur in Südamerika, sondern auch in mittelamerikanischen Ländern, wie Nicaragua, Costa-Rica, Panama usw., gibt es herrliche bunte Cichliden, die wir in Band II und III vorstellen werden.

Die meisten Cichliden betreiben sehr interessante Brutpflege, z.B. die „Höhlenbrüter", die in selbst geschaffenen Mulden oder vorhandenen Fels- oder Wurzelspalten ablaichen und diesen Brutplatz gegen Räubern oder vemeintlichen Eindringlingen aggressiv verteidigen.

Viele Arten sind „Maulbrüter", die ovophilen nehmen gleich die befruchteten Eier ins Maul und brüten sie aus, die larvophilen nehmen ihre „Kinder" erst, wenn sie ausgeschlüpft sind (Larven) bei Gefahr ins Maul und lassen sie wieder heraus, wenn die Gefahr vorüber ist.

Mit genauen Fundortangaben sollte man vorsichtig umgehen, denn in vielen Gebieten (z.B. Pantanal) , werden während jeder Regenzeit riesige Gebiete überschwemmt und verschiedene Flußsysteme miteinander verbunden, das ist etwa so, als wenn ein Fisch in Frankreich gefunden wird und nachher in Russland vorkäme.

Es sind eben enorm große Gebiete mit unvorstellbar großen Wasserflächen und einem gewaltigenFischreichtum und solange diese Biotope erhalten bleiben, und dafür müssen wir uns alle engagieren, kann dort niemals überfischt werden, die Fischer fangen wie eh und je mit Handnetzen ihre tägliche Mahlzeit und, wenn sie Lust haben, einige kleine Fische mehr, wofür sie etwas Geld bekommen, was sie auch dringend brauchen, denn es istmeist ihre einzige Einnahmequelle.

Außerdem gibt es inzwischen in fast allen Ursprungsländern große Freilandanlagen, in denen Speise- und Zierfische in großen Mengen und hervorragender Qualität für den Export gezüchtet werden.

Die meisten südamerikanischen Cichliden passen sich den Bedingungen im Aquarium leicht an, denn auch in der Natur müssen sie sich immer wieder auf die, je nach Jahreszeit, abweichenden Wasserwerte einstellen.

Die von uns angegebenen Zahlen beziehen sich auf Mittelwerte fürs Aquarium, schaffen Sie Ihren Pfleglingen die optimalen Bedingungen, in ihrem Wohlbefinden danken sie es Ihnen durch Farbenpracht und Vitalität, und wenn dann stolz die erste „Nachzucht" schwimmt ....,was kann es Schöneres für einen Hobby-Aquarianer geben.

SOUTH AMERICAN CICHLIDS I
What are cichlids?

Cichlids and their 'relatives' are found on all five continents - the majority of them and the most colourful in Africa and South America. In German, cichlids are known as 'Buntbarsche' -'ornamental perch'.

In Europe when people think of perch they automatically visualise a grey food fish. There are, however, here in Europe also very beautiful species, e.g. the European perch (Perca fluviatilis) or the pumpkin-seed perch (Lepomis) which originates from North America. These can be favourably compared with their cousins from other parts of the world. And in Africa and South America the larger species are also caught and eaten as food fish.

The more exotic coloured species are principally the ones kept in aquariums which is why they are known in German as "ornamental perch".

There is an unimaginably large number of perch types. There are the 'small fry' such as Elassoma evergladei which grow to no longer than 3 cm and there are the 'giants' which grow to a length of 100 cm.

They may be elongated such as the pike cichlids, may have a typical oval "fish shape" or an exotic shape or be plate-shaped such as the discus cichlids and scalars.

The cichlids belong principally to the more highly developed fish. They are mostly, although not exclusively, territorial by nature. Some are entirely 'peaceable' and are perfectly suited to a mixed aquarium, some are 'aggressive' to a greater or lesser extent and are intolerant of any other species and others are 'belligerent', looking for fights amongst their own kind.

All the above is especially true of the South American cichlids meaning that care should be taken when choosing which types to put together, and when deciding whether they are a good match and are able to coexist peacefully.

In addition to the South American species featured in this volume there are wonderful ornamental cichlids to be found in central American countries such as Nicaragua, Costa Rica, Panama etc. These will be presented in volumes II and III.

Most cichlids have very interesting breeding characteristics. For example, the 'cave breeders' which spawn in hollows which they create themselves, or in rock or root crevices. They then aggressively defend this spawning ground against raiders or supposed intruders.

Many species are 'mouth breeders'. The 'ovophiles' immediately suck up the fertilised eggs into their mouths and incubate them there. 'Larvophiles' take their young (larvae) into their mouths after they have hatched if danger is present, and release them when it has passed.

Details of a fish's place of origin should be regarded with caution as in many regions (e.g. Pantanal) vast areas are flooded every rainy season and many river systems become connected. This would be comparable to a fish occurring in France and then being found in Russia.

These are vast regions with unbelievably large stretches of water with an incredible variety of fish - as long as this biotope remains intact, a cause which all of us should support. They cannot be overfished as the fishermen there use the same methods they have for centuries of fishing for their daily food with hand nets. If they have time they catch a few more and sell them for money which they desperately need, as this is often their sole source of income.

In most of these native countries there are also large outdoor farms where large numbers of high quality edible and ornamental fish are bred for export.

Most South American cichlids adapt easily to the conditions in the aquarium as in their natural environment they have to accommodate the changing water values with each new season.

The figures we give refer to the mean aquarium values. If you provide your fish with optimum conditions then you will be rewarded with a wonderful display of colours and a high level of vitality. And for an aquarium enthusiast, there is nothing to beat the feeling when the next generation of successfully-bred cichlids starts swimming.

Lieber Leser,

sollten Sie im Besitz kleiner oder großer Dia- oder Foto-Sammlungen sein, bitte setzen Sie sich mit uns in Verbindung. Wir suchen für unsere nächsten Bücher immer gute Bilder von allen Fisch-Arten und besonders schönen Aquarien, und würden Ihre Bilder, natürlich gegen eine angemessene Benutzungsgebühr, gerne veröffentlichen.

To our readers,

if you are in possession of either a small or large collection of slides or photographs please contact us. For our upcoming books we are always on the lookout for good pictures of all types of fish and also for attractive aquariums. We would like to publish your photographs - obviously for a suitable charge.

S13703-4 Cichla sp. ARAGUAIA III ADULT
ARAGUAIA - CICHLID, till today not described!
Tocantins, Brazil, W, 55-60 cm

S13703-2 Cichla sp. ARAGUAIA III   JUVENIL
ARAGUAIA - CICHLA till today not described!
Tocantins, Brazil, W, 55-60 cm

S13718-1 Cichla ocellaris JUVENIL
PIKE - CICHLID
North-Southamerica, W, 40-50cm

S13718-4 Cichla ocellaris   ADULT
PIKE - CICHLID
North-Southamerica, W, 40-50cm

S13725-4 Cichla cf. ocellaris "RIO PRETO"
PRETO - PIKE-CICHLID
middle course of Rio Negro, W, 40-50cm

S13735-3 Cichla orinocensis ADULT
ORINOCO - PIKE-CICHLID
Rio Orinoco, Venezuela/Columbia, W, 50-60cm

S13735-4 Cichla orinocensis ADULT
ORINOCO - PIKE-CICHLID
Rio Orinoco, Venezuela/Columbia, W, 50-60cm

S13735-5 Cichla orinocensis ADULT
ORINOCO - PIKE-CICHLID
Rio Orinoco, Venezuela/Columbia, W, 50-60cm

South America **Cichlids I**

S13745-2 Cichla cf. orinocensis "RIO INIRIDA" JUVENIL
INIRIDA - PIKE-CICHLID
Orinoco - region, W, 50-60cm

S13748-2 Cichla cf. orinocensis "RIO VAUPES" JUVENIL
VAUPES - PIKE-CICHLID
Orinoco - region, W, 50-60cm

S13755-2 Cichla sp. TAPAJOS I JUVENIL
TAPAJOS - CICHLA
Brazil, W, 40-50cm

S13758-1 Cichla temensis JUVENIL
TEMENSIS - PIKE-CICHLID
Brazil, Venezuela, Columbia, W, 60-70cm

S13758-2 Cichla temensis JUVENIL
TEMENSIS - PIKE-CICHLID
Brazil, Venezuela, Columbia, W, 60-70cm

S13758-4 Cichla temensis ADULT
TEMENSIS - PIKE-CICHLID
Brazil, Venezuela, Columbia, W, 60-70cm

S13760-2 Cichla cf. temensis "CARONI" JUVENIL
CARONI - CICHLA
Venezuela, W, 60-70cm

S13760-3 Cichla cf. temensis "CARONI" SEMIADULT + ADULT
CARONI - CICHLA
Venezuela, W, 60-70cm

South America Cichlids I

© A.C.S. Glaser GmbH

S13762-3 Cichla cf. temensis "ARAGUAIA" SEMIADULT
ARAGUAIA - CICHLA
Araguaia-region, Brazil, W, 60-70cm

S13765-3 Cichla cf. temensis "VAUPES" SEMIADULT
VAUPES - CICHLA
Rio Vaupes, Orinoco-region, W, 60-70cm

S13763-2 Cichla cf. temensis "CAURA" JUVENIL
CAURA - CICHLA
Venezuela, W, 60-70cm

S13770-2 Cichla sp. XINGU II JUVENIL
XINGU II - CICHLA
Brazil, W, 40-50cm

S13770-4 Cichla sp. XINGU II ADULT
XINGU II - CICHLA
Brazil, W, 40-50cm

S23310-2 Crenicichla acutirostris JUVENIL
ACUTIS - CRENICICHLA
Rio Tapajos, North-East - Brazil, W, 26-32cm

S23310-3 Crenicichla acutirostris MALE
ACUTIS - CRENICICHLA
Rio Tapajos, North-East - Brazil, W, 26-32cm

S23310-3 Crenicichla acutirostris FEMALE
ACUTIS - CRENICICHLA
Rio Tapajos, North-East - Brazil, W, 26-30cm

South America **Cichlids I**

9

S23315-4 Crenicichla albopunctata MALE
SPOTTED - CRENICICHLA
Fr.Guayana + Surinam, W, 18-20cm

S23315-4 Crenicichla albopunctata FEMALE
SPOTTED - CRENICICHLA
Fr.Guayana + Surinam, W, 12-15cm

S23320-4 Crenicichla sp. ALTA MALE
ALTA - CRENICICHLA
Cuyuni-River-System, Venezu.+Guayana, W, 18-20cm

S23320-4 Crenicichla sp. ALTA FEMALE
ALTA - CRENICICHLA
Cuyuni-River-System, Venezu.+Guayana, W, 12-15cm

S23325-5 Crenicichla anthurus MALE
ANTHURUS - CRENICICHLA
Peru, Ecuador, West-Columbia, W, 18-20cm

S23325-5 Crenicichla anthurus FEMALE
ANTHURUS - CRENICICHLA
Peru, Ecuador, West-Columbia, W, 12-15cm

S23325-4 Crenicichla anthurus PAIR
ANTHURUS - CRENICICHLA
Peru, Ecuador, West-Columbia, W,

S23330-4 Crenicichla sp. APPROUAGE
APPROUAGE - CRENICICHLA
French - Guayana, W, 15-20cm

South America Cichlids I © A.C.S. Glaser GmbH

S23335-4 Crenicichla sp. ARAPIUNS   FEMALE
ARAPIUNS - CRENICICHLA
Rio Arapiuns, Tapajos-System, Brazil, W, 12-15cm

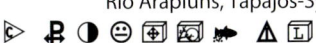

S23340-4 Crenicichla sp. ATABAPO   MALE
ATABAPO - CRENICICHLA
Rio Atabapo, Columbia, W, 25-30cm

S23340-4 Crenicichla sp. ATABAPO   FEMALE
ATABAPO - CRENICICHLA
Rio Atabapo, Columbia, W, 24-28cm

S23350-3 Crenicichla sp. BELEM
BELEM - PIKE-CICHLID
Brazil, W, 20-25cm

S23355-4 Crenicichla sp. BELLY-CRAWLER   MALE
BELLY-CRAWLER - CRENICICHLA
Rio Meta, Columbia, W, 20-22cm

S23355-4 Crenicichla sp. BELLY-CRAWLER   FEMALE
BELLY-CRAWLER - CRENICICHLA
Rio Meta, Columbia, W, 16-18cm

S23360-3 Crenicichla sp. BOCON   MALE
BOCON - CRENICICHLA
Rio Inirida - System, Columbia, W, 16-18cm

S23360-3 Crenicichla sp. BOCON   FEMALE
BOCON - CRENICICHLA
Rio Inirida - System, Columbia, W, 12-14cm

South America **Cichlids I**

CICHLA OCELLARIS

CICHLA OCELLARIS

CRENICICHLA SP. "TAPAJOS I"   FEMALE

S23370-4 Crenicichla cametana MALE
TOCANTINS - CRENICICHLA
Rio Tocantins, North-East - Brazil, W, 22-25cm

S23370-4 Crenicichla cametana FEMALE
TOCANTINS - CRENICICHLA
Rio Tocantins, North-East - Brazil, W, 18-20cm

S23378-4 Crenicichla cardiostigma (pterogramma?) MALE
PTEROGRAMMA - CRENICICHLA
upper course at Rio-Branco, North-Brazil, W, 18-20cm

S23378-4 Crenicichla cardiostigma (pterogramma?) Female
PTEROGRAMMA - CRENICICHLA
upper course at Rio-Branco, North-Brazil, W, 12-15cm

S23375-5 Crenicichla cardiostigma Male
RIO-BRANCO - CRENICICHLA
Rio Branco, Brazil, W, 18-20cm

S23375-5 Crenicichla cardiostigma FEMALE
RIO-BRANCO - CRENICICHLA
Rio Branco, Brazil, W, 16-18cm

S23380-4 Crenicichla sp. CASIQUIARE FEMALE
CASIQUIARE - CRENICICHLA
upper-course of Orinoco, Columbia, W, 12-15cm

S23385-2 Crenicichla cincta JUVENIL
CINCTA - CRENICICHLA
Amazonas, Brazil, W, 35-40cm

14 South America Cichlids I © A.C.S. Glaser GmbH

S23390-5 Crenicichla compressiceps   MALE
COMPRESS - CRENICICHLA
Rio Tocantins, North-East-Brazil, W, 7-8cm

S23390-5 Crenicichla compressiceps   FEMALE
COMPRESS - CRENICICHLA
Rio Tocantins, North-East-Brazil, W, 5-6cm

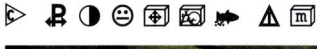

S23395-4 Crenicichla cyanonotus   MALE
CYANO - CRENICICHLA
Amazonas-System, Brazil, W, 22-25cm

S23395-4 Crenicichla cyanonotus   FEMALE
CYANO - CRENICICHLA
Amazonas-System, Brazil, W, 18-20cm

S23400-2 Crenicichla cyclostoma   JUVENIL
CYCLOS - CRENICICHLA
Rio Tocantins, North-East-Brazil, W, 14-16cm

S23400-5 Crenicichla cyclostoma   MALE
CYCLOS - CRENICICHLA
Rio Tocantins, North-East-Brazil, W, 14-16cm

S23400-5 Crenicichla cyclostoma   FEMALE
CYCLOS - CRENICICHLA
Rio Tocantins, North-East-Brazil, W, 11-13cm

S23415-4 Crenicichla dorsocellata   FEMALE
CAMPOS - CRENICICHLA
Campos, East-Brazil, W, 18-20cm

South America **Cichlids I**   15

S23425-4 Crenicichla edithae MALE
EDITH - CRENICICHLA
Rio Paraguaya, Brazil, W, 18-20cm

S23425-4 Crenicichla edithae FEMALE
EDITH - CRENICICHLA
Rio Paraguaya, Brazil, W, 14-16cm

S23435-4 Crenicichla frenata MALE
TRINIDAD - CRENICICHLA
Trinidad, W, 18-20cm

S23435-4 Crenicichla frenata FEMALE
TRINIDAD - CRENICICHLA
Trinidad, W, 13-15cm

S23440-4 Crenicichla frenata var. MALE
FRENATA - CRENICICHLA
Columbia, W, 18-20cm

S23440-4 Crenicichla frenata var. FEMALE
FRENATA - CRENICICHLA
Columbia, W, 13-15cm

S23450-4 Crenicichla geayi MALE
GEAYI - CRENICICHLA
Orinoco-System, Columbia, W, 18-20cm

S23450-4 Crenicichla geayi FEMALE
GEAYI - CRENICICHLA
Orinoco-System, Columbia, W, 14-16cm

South America Cichlids I   © A.C.S. Glaser GmbH

S23455-4 Crenicichla cf. geayi MALE
BRIGHT-BLUE - CRENICICHLA
Orinoco / Venezuela, W, 15-17cm

S23455-4 Crenicichla cf. geayi FEMALE
BRIGHT-BLUE - CRENICICHLA
Orinoco / Venezuela, W, 12-15cm

S23455-5 Crenicichla cf. geayi PAIR at breeding-care
BRIGHT-BLUE - CRENICICHLA
Orinoco / Venezuela, W,

S23465-4 Crenicichla sp. GUANIAMO MALE
GUANIAMO - CRENICICHLA
upper Orinoco, Venezuela, W, 18-20cm

S23465-4 Crenicichla sp. GUANIAMO FEMALE
GUANIAMO - CRENICICHLA
upper Orinoco, Venezuela, W, 16-18cm

S23470-4 Crenicichla sp. GUARIQUITO MALE
GUARIQUITO - CRENICICHLA
Rio Guariquito, Venezuela, W, 18-20cm

S23470-4 Crenicichla sp. GUARIQUITO FEMALE
GUARIQUITO - CRENICICHLA
Rio Guariquito, Venezuela, W, 13-15cm

S23475-4 Crenicichla sp. var. GUARIQUITO MALE
AYACUCHO - CRENICICHLA
Puerto Ayacucho, Columbia, W, 18-20cm

South America Cichlids I   17

S23475-4 Crenicichla sp. var. GUARIQUITO FEMALE
AYACUCHO - CRENICICHLA
Puerto Ayacucho, Columbia, W, 13-15cm

S23480-5 Crenicichla sp. GUAYANA MALE
DEMERARA - CRENICICHLA
Demerara-River, Guayana, W, 18-20cm

S23480-5 Crenicichla sp. GUAYANA FEMALE
DEMERARA - CRENICICHLA
Demerara-River, Guayana, W, 13-18cm

S23490-4 Crenicichla heckeli MALE
HECKEL - CRENICICHLA
Rio Trombetas, North-East-Brazil, W, 6-7cm

S23490-4 Crenicichla heckeli FEMALE
HECKEL - CRENICICHLA
Rio Trombetas, North-East-Brazil, W, 4-5cm

S23500-4 Crenicichla sp. INIRIDA I MALE
INIRIDA I - CRENICICHLA
Rio Inirida, Columbia, W, 18-20cm

S23500-4 Crenicichla sp. INIRIDA I FEMALE
INIRIDA I - CRENICICHLA
Rio Inirida, Columbia, W, 16-18CM

S23505-4 Crenicichla sp. INIRIDA II FEMALE
INIRIDA II - CRENICICHLA
Rio Inirida, Columbia, W, M.14-16cm, F.12-14cm

South America **Cichlids I**   © A.C.S. Glaser GmbH

S23510-4 Crenicichla sp. INIRIDA III "RED-SPOT" MALE
INIRIDA III - RED-SPOT - CRENICICHLA
Rio Inirida, Columbia, W, 15-17cm

S23510-4 Crenicichla sp. INIRIDA III "RED-SPOT" FEMALE
INIRIDA III - RED-SPOT - CRENICICHLA
Rio Inirida, Columbia, W, 13-15cm

S23520-4 Crenicichla sp. JABUTI MALE
JABUTI - CRENICICHLA
Rio Tapajos, Brazil, W, 16-18cm

S23520-4 Crenicichla sp. JABUTI FEMALE
JABUTI - CRENICICHLA
Rio Tapajos, Brazil, W, 14-16cm

S23525-5 Crenicichla jaguarensis FEMALE
JAGUAR - CRENICICHLA
Rio Parana, East-Brazil, W, M.20-22cm, F.18-20cm

S23530-2 Crenicichla jegui JUVENIL
JEGUI - CRENICICHLA
Rio Tocantins, North-East-Brazil, W, 26-28cm

S23530-4 Crenicichla jegui MALE
JEGUI - CRENICICHLA
Rio Tocantins, North-East-Brazil, W, 26-28cm

S23530-4 Crenicichla jegui FEMALE
JEGUI - CRENICICHLA
Rio Tocantins, North-East-Brazil, W, 20-22cm

South America Cichlids I

S23535-5 Crenicichla sp. jegui MALE
NORTH-EAST - CRENICICHLA MALE 23-25cm
Rio Tocantins, North-East-Brazil, W, FEMALE 18-20cm

S23540-4 Crenicichla johanna JUVENIL
JOHANNA - CRENICICHLA
Amazonas-Area, Brazil, W, 35-40cm

S23540-4 Crenicichla johanna MALE
JOHANNA - CRENICICHLA
Amazonas-Area, Brazil, W, 35-40cm

S23540-4 Crenicichla johanna FEMALE
JOHANNA - CRENICICHLA
Amazonas-Area, Brazil, W, 30-35cm

S23545-4 Crenicichla cf. johanna var. FEMALE
JOHANNA-VAR: - CRENICICHLA
Orinoco, Columbia, W, M.34-38cm + F.28-32cm

S23560-4 Crenicichla labrina MALE
LABRINA - CRENICICHLA
Rio-Tocantins-System, North-East-Brazil, W, 18-20cm

S23560-4 Crenicichla labrina FEMALE
LABRINA - CRENICICHLA
Rio-Tocantins-System, North-East-Brazil, W, 12-15cm

S23565-5 Crenicichla sp. labrina MALE
AMAZONAS - CRENICICHLA M. 18-20cm
down below Amazonas, N/E-Brazil, W, F. 12-15cm

South America Cichlids I   © A.C.S. Glaser GmbH

CRENICICHLA COMPRESSICEPS MALE

S23570-2 Crenicichla lenticulata   JUVENIL
LENTICULATA - CRENICICHLA
Rio Negro/Orinoco, N-Brazil + Columbia, W, 35-38cm

S23570-4 Crenicichla lenticulata   MALE
LENTICULATA - CRENICICHLA
Rio Negro/Orinoco, N-Brazil + Columbia, W, 35-38cm

S23570-4 Crenicichla lenticulata   FEMALE
LENTICULATA - CRENICICHLA
Rio Negro/Orinoco, N-Brazil + Columbia, W, 34-36cm

S23575-4 Crenicichla lepidota   MALE
LEPIDOTA - CRENICICHLA   not too agressiv !!
Rio Paraguaya, Brazil, W, 14-16cm

S23575-4 Crenicichla lepidota   FEMALE
LEPIDOTA - CRENICICHLA   not too agressiv !!
Rio Paraguaya, Brazil, W, 13-14cm

S23580-4 Crenicichla lepidota r e a l MALE
REAL LEPIDOTA - CRENICICHLA
down course at Rio Guapore, N/E-Brazil, W, 13-15cm

S23580-4 Crenicichla lepidota r e a l FEMALE
REAL LEPIDOTA - CRENICICHLA
down course at Rio Guapore, N/E-Brazil, W, 10-12cm

S23590-4 Crenicichla lucius MALE
LUCIUS - CRENICICHLA
upper Amazonas, Peru + W-Brazil, W, 18-20cm

South America Cichlids I   © A.C.S. Glaser GmbH

S23590-4 Crenicichla lucius   FEMALE
LUCIUS - CRENICICHLA   Male   18-20cm
upper Amazonas, Peru + W-Brazil, W, Female 13-15cm

S23595-2 Crenicichla lugubris   JUVENIL
LUGU - CRENICICHLA
Amazonas-System, Brazil, W, 35-38cm

S23595-4 Crenicichla lugubris   MALE
LUGU - CRENICICHLA
Amazonas-System, Brazil, W, 35-38cm

S23595-4 Crenicichla lugubris   FEMALE
LUGU - CRENICICHLA
Amazonas-System, Brazil, W, 33-35cm

S23596-5 Crenicichla lugubris var.   MALE
RED-MOUTH - CRENICICHLA
Rio Araguaya, N/E-Brazil, W, 35-38cm

S23596-5 Crenicichla lugubris var.   FEMALE
RED-MOUTH - CRENICICHLA
Rio Araguaya, N/E-Brazil, W, 33-35cm

S23600-2 Crenicichla cf. lugubris "ATABAPO" JUVENIL
ATABAPO - CRENICICHLA
Rio Atabapo, Columbia, W, 31-35cm

S23600-5 Crenicichla cf. lugubris "ATABAPO" FEMALE ADULT
ATABAPO - CRENICICHLA
Rio Atabapo, Columbia, W, 30-33cm

South America Cichlids I   23

CRENICICHLA LUGUBRIS   FEMALE

South America Cichlids I

CRENICICHLA NOTOPHTHALMUS   FEMALE

S23610-3 Crenicichla macrophthalma FEMALE
MACRO - CRENICICHLA
Rio Negro, North-Brazil, W, M.23-25cm, F.20-22cm

S23615-4 Crenicichla sp. MAICURU FEMALE
MAICURU - CRENICICHLA   M.18-20cm
down course at Rio Maicuru, N/E-Brazil, W, F.13-15cm

S23620-2 Crenicichla sp. MANAUS JUVENIL
MANAUS - CRENICICHLA
Rio Negro - System, North-Brazil, W,

S23620-4 Crenicichla sp. MANAUS MALE
MANAUS - CRENICICHLA
Rio Negro - System, North-Brazil, W, 16-18cm

S23620-4 Crenicichla sp. MANAUS FEMALE
MANAUS - CRENICICHLA
Rio Negro - System, North-Brazil, W, 12-14cm

S23625-1 Crenicichla marmorata JUVENIL
MARBLE - CRENICICHLA
Rio Tapajos, down Amazonas-System, W,

S23625-2 Crenicichla marmorata JUVENIL MALE
MARBLE - CRENICICHLA
Rio Tapajos, down Amazonas-System, W, 32-38cm

S23625-2 Crenicichla marmorata JUVENIL FEMALE
MARBLE - CRENICICHLA
Rio Tapajos, down Amazonas-System, W, 30-36cm

26 South America Cichlids I © A.C.S. Glaser GmbH

S23625-4 Crenicichla marmorata   MALE
MARBLE - CRENICICHLA
Rio Tapajos, down Amazonas-System, W, 32-38cm

S23625-4 Crenicichla marmorata   FEMALE
MARBLE - CRENICICHLA
Rio Tapajos, down Amazonas-System, W, 30-36cm

S23630-4 Crenicichla sp. marmorata   MALE
MARMORATA - CRENICICHLA
Rio Tapajos, N/E-Brazil, W, M.33-35cm, F.30-33cm

S23635-5 Crenicichla sp. "MATO-GROSSO"   MALE
MATO-GROSSO - CRENICICHLA
Mato-Grosso, Brazil, W, 18-20cm

S23635-5 Crenicichla sp. "MATO-GROSSO"   FEMALE
MATO-GROSSO - CRENICICHLA
Mato-Grosso, Brazil, W, 16-18cm

S23635-5 Crenicichla sp. "MATO-GROSSO"   PAIR
MATO-GROSSO - CRENICICHLA
Mato-Grosso, Brazil, W,

S23640-2 Crenicichla sp. menezesi   JUVENIL
MENEZESI - CRENICICHLA
Pernambuco, East-Brazil, W,

S23640-5 Crenicichla sp. menezesi   MALE
MENEZESI - CRENICICHLA
Pernambuco, East-Brazil, W, 14-16cm

South America Cichlids I   27

S23640-5 Crenicichla sp. menezesi FEMALE
MENEZESI - CRENICICHLA
Pernambuco, East-Brazil, W, 8-10cm

S23645-4 Crenicichla multispinosa MALE
MULTI - CRENICICHLA M.30-32cm
Rio Maronii, Fr.-Guayana + Surinam, W, F.28-30cm

S23655-3 Crenicichla sp. nothophthalmus MALE
NEGRO - CRENICICHLA
Orinoco-System, Columbia/Venezuela, W, 12-14cm

S23655-3 Crenicichla sp. nothophthalmus FEMALE
NEGRO - CRENICICHLA
Orinoco-System, Columbia/Venezuela, W, 7-9cm

S23660-4 Crenicichla notophthalmus MALE
NORTH - CRENICICHLA
Rio Negro, North-Brazil, W, 10-12cm

S23660-4 Crenicichla notophthalmus FEMALE
NORTH - CRENICICHLA
Rio Negro, North-Brazil, W, 7-8cm

S23663-3 Crenicichla cf. notophthalmus MALE
BIG-POINT - CRENICICHLA
Brazil, W, 9-11cm

S23663-3 Crenicichla cf. notophthalmus FEMALE
BIG-POINT - CRENICICHLA
Brazil, W, 6-8cm

South America Cichlids I © A.C.S. Glaser GmbH

1. **Südamerikanisches Biotop**

   South American Biotope

2. **Überall sind beliebte Aquariumfische auch begehrte Speisefische.**

   Everywhere popular aquarium fish are also sought-after food fish

CRENICICHLA PHAIOSPILUS

S23664-4 Crenicichla sp. notophthalmus "UAUPES" MALE
UAUPES - CRENICICHLA
Rio Uaupes, W, 10-12cm

S23664-4 Crenicichla sp. notophthalmus "UAUPES" FEMALE
UAUPES - CRENICICHLA
Rio Uaupes, W, 8-9cm

S23670-3 Crenicichla sp. "ORANGE" - XINGU/BELEM
ORANGE-XINGU - CRENICICHLA
Belem, Brazil, W, 16-18cm

S23680-4 Crenicichla sp. ORINOCO MALE
MANY-SPOT - CRENICICHLA
Orinoco, Venezuela + Columbia, W, 18-20cm

S23680-4 Crenicichla sp. ORINOCO FEMALE
MANY-SPOT - CRENICICHLA
Orinoco, Venezuela + Columbia, W, 13-15cm

S23675-4 Crenicichla ORINOCO MALE
ORINOCO - CRENICICHLA
Orinoco/Caroni, Venezuela, W, 11-13cm

S23675-4 Crenicichla ORINOCO FEMALE
ORINOCO - CRENICICHLA
Orinoco/Caroni, Venezuela, W, 7-9cm

S23690-5 Crenicichla sp. PACAYA MALE
PACAYA - CRENICICHLA   M.18-20cm
Rio Pacaya, down course at Anazonas, W, F. 15-17cm

South America **Cichlids I**   31

S23695-4 Crenicichla percna   MALE
PERCNA - CRENICICHLA
Rio Xingu, N/E.Brazil, W, 32-36cm

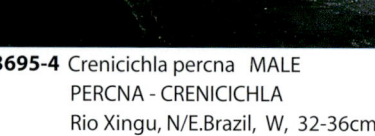

S23695-4 Crenicichla percna   FEMALE
PERCNA - CRENICICHLA
Rio Xingu, N/E.Brazil, W, 30-34cm

S23705-4 Crenicichla sp. PERNAMBUCO   MALE
PERNAMBUCO - CRENICICHLA
Pernambuco, Brazil, W, 11-13cm

S23705-4 Crenicichla sp. PERNAMBUCO   FEMALE
PERNAMBUCO - CRENICICHLA
Pernambuco, Brazil, W, 6-8cm

S23705-5 Crenicichla sp. PERNAMBUCO   PAIR
PERNAMBUCO - CRENICICHLA
Pernambuco, Brazil, W,

S23710-1 Crenicichla phaiospilus Juvenil
SPILUS - CRENICICHLA
Rio Xingu, N/E-Brazil, W,

S23710-2 Crenicichla phaiospilus   JUVENIL
SPILUS - CRENICICHLA
Rio Xingu, N/E-Brazil, W,

S23710-3 Crenicichla phaiospilus
SPILUS - CRENICICHLA   MALE   33-36cm
Rio Xingu, N/E-Brazil, W, FEMALE 28-32cm

South America Cichlids I   © A.C.S. Glaser GmbH

S23715-4 Crenicichla sp. PINDARE
PINDARE - CRENICICHLA   MALE   18-20cm
Rio Pindare, Brazil, W,   FEMALE 14-16cm

S23720-4 Crenicichla proteus   MALE
PERU - CRENICICHLA
Rio Ucayali - System, Peru, W, 16-18cm

S23720-4 Crenicichla proteus   FEMALE
PERU - CRENICICHLA
Rio Ucayali - System, Peru, W, 12-14cm

S23725-4 Crenicichla punctata   MALE
PUNCTATA - CRENICICHLA
Rio dos Sinos, South-East-Brazil, 22-25cm

S23725-4 Crenicichla punctata   FEMALE
PUNCTATA - CRENICICHLA
Rio dos Sinos, South-East-Brazil, 18-20cm

S23735-4 Crenicichla sp. "RED-BELLY"
RED-BELLY - CRENICICHLA
Brazil, W, Male 18-20cm + Female 14-16cm

S23735-4 Crenicichla sp. "RED-BELLY"
RED-BELLY - CRENICICHLA
Brazil, W, Male 18-20cm + Femal 14-16cm

S23735-5 Crenicichla sp. "RED-BELLY"
RED-BELLY - CRENICICHLA
Brazil, W, Male 18-20cm + Femal 14-16cm

South America Cichlids I

CRENICICHLA REGANI "RIO NEGRO"

Crenicichla regani "RIO das MORTES"

S23737-4 Crenicichla regani "das MORTES"
MORTES - CRENICICHLA
Araguaia, Brazil, W, Male 13cm + FEMALE 9cm

S23737-4 Crenicichla regani "das MORTES"
MORTES - CRENICICHLA
Araguaia, Brazil, W, Male 13cm + FEMALE 9cm

S23738-4 Crenicichla regani "RIO-GUAMA"
GUAMA - CRENICICHLA
North-East-Brazil, W, Male 13cm + Female 9cm

S23738-4 Crenicichla regani "RIO-GUAMA"
GUAMA - CRENICICHLA
North-East-Brazil, W, Male 13cm + Female 9cm

S23739-3 Crenicichla regani "RIO-ACARA"
ACARA - CRENICICHLA
North-East-Brazil, W, Male 13cm + Female 9cm

S23739-3 Crenicichla regani "RIO-ACARA"
ACARA - CRENICICHLA
North-East-Brazil, W, Male 13cm + Female 9cm

S23740-4 Crenicichla regani
REGANI - CRENICICHLA
Rio Capari, Tapajos, W, Male 13cm + Female 9cm

S23740-4 Crenicichla regani
REGANI - CRENICICHLA
Rio Capari, Tapajos, W, Male 13cm + Female 9cm

South America Cichlids I

S23741-4 Crenicichla regani "RIO-NEGRO"
RIO-NEGRO - CRENICICHLA
North-Brazil, Tapajos, W, Male 13cm + Female 9cm

S23741-4 Crenicichla regani "RIO-NEGRO"
RIO-NEGRO - CRENICICHLA
North-Brazil, Tapajos, W, Male 13cm + Female 9cm

S23742-3 Crenicichla regani "TEFE"
TEFE - CRENICICHLA
West-Brazil, Tapajos, W, Male 13cm + Female 9cm

S23742-3 Crenicichla regani "TEFE"
TEFE - CRENICICHLA
West-Brazil, Tapajos, W, Male 13cm + Female 9cm

S23743-4 Crenicichla regani "TROMBETAS"
TROMBETAS - CRENICICHLA
North-Brazil, W, Male 13cm + Female 9cm

S23743-4 Crenicichla regani "TROMBETAS"
TROMBETAS - CRENICICHLA
North-Brazil, W, Male 13cm + Female 9cm

S23744-4 Crenicichla regani "GUAPORE"
GUAPORE - CRENICICHLA
West-Brazil, W, Male 13cm + Female 9cm

S23744-4 Crenicichla regani "GUAPORE"
GUAPORE - CRENICICHLA
West-Brazil, W, Male 13cm + Female 9cm

South America Cichlids I   © A.C.S. Glaser GmbH

S23745-4 Crenicichla regani "TAPAJOS"
TAPAJOS - CRENICICHLA
Brazil, W, Male 13cm + Female 9cm

S23745-4 Crenicichla regani "TAPAJOS"
TAPAJOS - CRENICICHLA
Brazil, W, Male 13cm + Female 9cm

S23748-4 Crenicichla sp. regani MALE
BLACK-LINE - CRENICICHLA
Rio Xingu, N/E-Brazil, W, 8-9cm

S23748-4 Crenicichla sp. regani FEMALE
BLACK-LINE - CRENICICHLA
Rio Xingu, N/E-Brazil, W, 5-6cm

S23750-4 Crenicichla reticulata MALE
RETICULATA - CRENICICHLA
Amazonas - System, Brazil, W, 22-25cm

S23750-4 Crenicichla reticulata FEMALE
RETICULATA - CRENICICHLA
Amazonas - System, Brazil, W, 18-20cm

S23755-4 Crenicichla sp. RIO-BRANCO FEMALE
RIO-BRANCO - CRENICICHLA
Rio Branco, Brazil, W, M.10-12cm + F.8-9cm

S23760-4 Crenicichla saxatilis "SURINAM" MALE
SAXA-SURINAM - CRENICICHLA
Guayana/Surinam, W, 18-20cm

South America Cichlids I

S23760-4 Crenicichla saxatilis "SURINAM" FEMALE
SAXA-SURINAM - CRENICICHLA
Guayana/Surinam, W, 13-15cm

S23758-4 Crenicichla saxatilis MALE
SAXA - CRENICICHLA
Fr.-Guayana, W, 18-20cm

S23758-4 Crenicichla saxatilis FEMALE
SAXA - CRENICICHLA
Fr.-Guayana, W, 15-18cm

S23765-2 Crenicichla cf. saxatilis JUVENIL
PEARL - CRENICICHLA
Trinidad, W,

S23765-4 Crenicichla cf. saxatilis MALE
PEARL - CRENICICHLA
Trinidad, W, 18-20cm

S23765-4 Crenicichla cf. saxatilis FEMALE
PEARL - CRENICICHLA
Trinidad, W, 16-18cm

S23775-4 Crenicichla sp. saxatilis MALE
BIG-POINT - CRENICICHLA
Belem, N/E-Brazil, W, 18-20cm

S23775-4 Crenicichla sp. saxatilis FEMALE
BIG-POINT - CRENICICHLA
Belem, N/E-Brazil, W, 13-15cm

38 South America Cichlids I © A.C.S. Glaser GmbH

CRENICICHLA SEMIFASCIATA FEMALE

South America **Cichlids I**

S23780-4 Crenicichla semifasciata MALE
SEMI - CRENICICHLA
Rio Paraguaya, Paraguay, W, 22-25cm

S23780-4 Crenicichla semifasciata FEMALE
SEMI - CRENICICHLA
Rio Paraguaya, Paraguay, W, 18-20cm

S23785-4 Crenicichla sp. SINOP MALE
SINOP - CRENICICHLA
Mato-Grosso, Brazil, W, 18-20cm

S23785-4 Crenicichla sp. SINOP FEMALE
SINOP - CRENICICHLA
Mato-Grosso, Brazil, W, 16-18cm

S23790-4 Crenicichla stocki MALE
STOCKI - CRENICICHLA
Rio Tocantins, N/E-Brazil, W, 22-25cm

S23790-4 Crenicichla stocki FEMALE
STOCKI - CRENICICHLA
Rio Tocantins, N/E-Brazil, W, 18-20cm

S23795-2 Crenicichla strigata JUVENIL
STRIGATA - CRENICICHLA
Rio Tapajos, Tocantins, Guama, N/E-Brazil, W,

S23795-4 Crenicichla strigata FEMALE
STRIGATA - CRENICICHLA   M.35-38cm + F.30-35cm
Rio Tapajos, Tocantins, Guama, N/E-Brazil, W,

South America Cichlids I © A.C.S. Glaser GmbH

S23800-2 Crenicichla sp. SURINAM JUVENIL
SURINAM - CRENICICHLA
Surinam, W, Male 22-25cm + Female 18-20cm

S23800-3 Crenicichla sp. SURINAM MALE
SURINAM - CRENICICHLA
Surinam, W, Male 22-25cm + Female 18-20cm

S23805-4 Crenicichla sveni MALE
META - CRENICICHLA
Rio-Meta-System, Columbia, W, 18-20cm

S23805-4 Crenicichla sveni FEMALE
META - CRENICICHLA
Rio-Meta-System, Columbia, W, 13-15cm

S23810-3 Crenicichla cf. sveni MALE
GAITAN - CRENICICHLA
Gaitan (Meta), Columbia, W, 18-20cm

S23810-3 Crenicichla cf. sveni FEMALE
GAITAN - CRENICICHLA
Gaitan (Meta), Columbia, W, 16-18cm

S23810-5 Crenicichla cf. sveni PAIR
GAITAN - CRENICICHLA
Gaitan (Meta), Columbia, W,

S23820-4 Crenicichla sp. TAPAJOS I FEMALE
TAPAJOS I - CRENICICHLA
Rio Tapajos, W, Male 33-35cm + Female 28-30cm

South America **Cichlids I**

CRENICICHLA SP. XINGU I   FEMALE

S23825-2 Crenicichla sp. TAPAJOS   JUVENIL
TAPAJOS - CRENICICHLA
Rio Tapajos, W, Male 33-35cm + Female 28-30cm

S23830-4 Crenicichla ternetzi   MALE
TERNETZI - CRENICICHLA   Male   28-32cm
Oyapock-River, Fr.-Guayana,, W,   Female 25-28cm

S23835-2 Crenicichla tigrina   JUVENIL
TIGRINA - CRENICICHLA   Male   32-35cm
Rio Trombetas, N/E-Brazil, W,   Female 30-33cm

S23845-5 Crenicichla urosema   Male
UROSEMA - CRENICICHLA
Rio Tapajos, N/E-Brazil, W, 8-10cm

S23845-5 Crenicichla urosema   FEMALE
UROSEMA - CRENICICHLA
Rio Tapajos, N/E-Brazil, W, 6-8cm

S23850-2 Crenicichla sp. URUBAXI   JUVENIL
URUBAXI - CRENICICHLA   Male   18-20cm
Amazonas, Brazil, W,   Female 15-17cm

S23850-4 Crenicichla sp. URUBAXI   MALE
URUBAXI - CRENICICHLA   Male   18-20cm
Amazonas, Brazil, W,   Female 15-17cm

S23860-2 Crenicichla sp. VENEZUELA (befor strigata)   JUVENIL
VENEZUELA - CRENICICHLA
Orinoco-System, Columbia + Venezuela

South America **Cichlids I**   43

S23860-3 Crenicichla sp. VENEZUELA (befor strigata) PAIR
VENEZUELA - CRENICICHLA   up: Male + down: Female
Orinoco-System, Columbia + Venezuela

S23863-2 Crenicichla sp. VENEZUELA   JUVENIL
JUMBO - CRENICICHLA   Male   35-40cm
Orinoco-Caroni, Venezuela, W,   Female 33-38cm

S23863-4 Crenicichla sp. VENEZUELA   PAIR
JUMBO - CRENICICHLA   Male   35-40cm
Orinoco-Caroni, Venezuela, W,   Female 33-38cm

S23865-1 Crenicichla sp. VENEZUELA - zwerg/dwarf   JUVENIL
DWARF - CRENICICHLA
Venezuela, W,   Male 15-18cm + Female 14-16cm

S23865-2 Crenicichla sp. VENEZUELA - zwerg/dwarf   JUVENIL
DWARF - CRENICICHLA
Venezuela, W,   Male 15-18cm + Female 14-16cm

S23865-3 Crenicichla sp. VENEZUELA - zwerg/dwarf   PAIR
DWARF - CRENICICHLA
Venezuela, W,   Male 15-18cm + Female 14-16cm

S23870-2 Crenicichla vittata   FEMALE - JUVENIL
VITTATA - CRENICICHLA   Male   30-34cm
Paraguay-Area, Rio Paraguaya, W,   Female 26-30cm

S23870-3 Crenicichla vittata   MALE
VITTATA - CRENICICHLA   Male   30-34cm
Paraguay-Area, Rio Paraguaya, W,   Female 26-30cm

South America **Cichlids I**   © A.C.S. Glaser GmbH

S23870-4 Crenicichla vittata FEMALE
VITTATA - CRENICICHLA Male 30-34cm
Paraguay-Area, Rio Paraguaya, W, Female 26-30cm

S23880-4 Crenicichla wallacii (?) MALE
WALLACII - CRENICICHLA
Rio Branco, North-Brazil, W, 11-12cm

S23880-4 Crenicichla wallacii (?) FEMALE
WALLACII - CRENICICHLA
Rio Branco, North-Brazil, W, 8-9cm

S23885-4 Crenicichla cf. wallacii (?) MALE
AYACUCHO - CRENICICHLA Male 12-14cm
Puerto Ayacucho, Venezuela, W, Female 10-12cm

S23895-2 Crenicichla sp. XINGU I JUVENIL
XINGU I - CRENICICHLA
Rio Xingu, N/E-Brazil, W, 35-38cm

S23895-4 Crenicichla sp. XINGU I MALE
XINGU I - CRENICICHLA
Rio Xingu, N/E-Brazil, W, 35-38cm

S23895-4 Crenicichla sp. XINGU I FEMALE
XINGU I - CRENICICHLA
Rio Xingu, N/E-Brazil, W, 30-32cm

S23900-2 Crenicichla sp. XINGU II JUVENIL
XINGU II - CRENICICHLA
Rio Xingu, N/E-Brazil

South America Cichlids I 45

S23900-4 Crenicichla sp. XINGU II   MALE
XINGU II - CRENICICHLA
Rio Xingu, N/E-Brazil, W, 34-36cm

S23900-4 Crenicichla sp. XINGU II   FEMALE
XINGU II - CRENICICHLA
Rio Xingu, N/E-Brazil, W, 32-34cm

S23905-2 Crenicichla sp. XINGU III   JUVENIL
XINGU III - CRENICICHLA
Rio Xingu, N/E-Brazil, W, M.34-36cm + F.32-34cm

S23905-4 Crenicichla sp. XINGU III   FEMALE
XINGU III - CRENICICHLA
Rio Xingu, N/E-Brazil, W, M.34-36cm + F.32-34cm

S23910-2 Crenicichla sp. XINGU-ORANGE   JUVENIL
ORANGE - CRENICICHLA
Rio Xingu, N/E-Brazil, W, M.30-33cm + F.28-30cm

S23910-3 Crenicichla sp. XINGU-ORANGE
ORANGE - CRENICICHLA
Rio Xingu, N/E-Brazil, W, M.30-33cm + F.28-30cm

Apistogramma agassizii RED - TAIL
Der, und alle anderen demnächst / all other coming soon
in "SOUTHAMERICAN - CICHLIDS II"

Apistogramma hongsloi II
Der, und alle anderen demnächst / all other coming soon
in "SOUTHAMERICAN - CICHLIDS II"

46   South America **Cichlids I**   © A.C.S. Glaser GmbH

GEOPHAGUS SP. "AREÕES"

S32105-4 Geophagus sp. ALTAMIRA
ALTAMIRA - GEOPHAGUS
Xingu, Brazil, W, 23-25cm

S32110-3 Geophagus altifrons
ALTIFRONS - GEOPHAGUS
Brazil, W, 22-24cm

S32115-4 Geophagus sp. altifrons "RIO - NEGRO"
RIO-NEGRO - ALTIFRONS
Rio Negro, Brazil, W, 23-25cm

S32115-4 Geophagus sp. altifrons "RIO - NEGRO"
RIO-NEGRO - ALTIFRONS
Rio Negro, Brazil, W, 23-25cm

S32120-3 Geophagus altifrons "TAPAJOS"
TAPAJOS - ALTIFRONS
Rio Tapajos, Brazil, W, 23-25cm

S32125-3 Geophagus altifrons "XINGU"
XINGU - ALTIFRONS
Rio Xingu, Brazil, W, 23-25cm

S32130-3 Geophagus cf. altifrons
ALTIFRONS - GEOPHAGUS
Rio Trombetas, N/E-Brazil, W, 23-25cm

S32135-3 Geophagus sp. ARAGUAIA
ARAGUAIA - GEOPGAGUS
Brazil, W, 20-22cm

S32140-3 Geophagus sp. AREÖES
AREÖES - GEOPHAGUS
Brazil, Araguaia, Tocantins, W, 15-17cm

S32140-4 Geophagus sp. AREÖES PAIR
AREÖES - GEOPHAGUS
Brazil, Araguaia, Tocantins, W, 15-17cm

S32143-2 Geophagus sp. AREÖES - 2-SPOT Juvenil
TWO-SPOT - GEOPHAGUS
Brazil, Araguaia, Tocantins, W, 15-17cm

S32143-2 Geophagus sp. AREÖES - 2-SPOT
TWO-SPOT - GEOPHAGUS
Brazil, Araguaia, Tocantins, W, 15-17cm

S32145-2 Geophagus argyrostictus JUVENIL
ARGY - GEOPHAGUS
Xingu, Brazil, W, 18-20cm

S32145-4 Geophagus argyrostictus MALE ADULT
ARGY - GEOPHAGUS
Xingu, Brazil, W, 18-20cm

S32145-4 Geophagus argyrostictus PAIR with Eggs
ARGY - GEOPHAGUS
Xingu, Brazil, W, 18-20cm

S32150-2 Geophagus australis
ARGENTINA - GEOPHAGUS
Argentina, W, 16-18cm

South America **Cichlids I** 49

GEOPHAGUS ARGYROSTICTUS

GEOPHAGUS ARGYROSTICTUS

S32160-4 Geophagus brasiliensis   MALE
BRAZIL - GEOPHAGUS
South-Brazil, W, 27-30cm

S32160-4 Geophagus brasiliensis   FEMALE
BRAZIL - GEOPHAGUS
South-Brazil, W, 27-30cm

S32160-4 Geophagus brasiliensis   PAIR
BRAZIL - GEOPHAGUS
South-Brazil, W, 27-30cm

S32165-4 Geophagus brasiliensis   "form A"
BRAZIL A - GEOPHAGUS
Brazil, W, 25-28cm

S32165-4 Geophagus brasiliensis   "form A"
BRAZIL A - GEOPHAGUS
Brazil, W, 25-28cm

S32170-4 Geophagus brasiliensis   "form B"
BRAZIL B - GEOPHAGUS
Brazil, W, 23-26cm

S32170-4 Geophagus brasiliensis   "form B"
BRAZIL B - GEOPHAGUS
Brazil, W, 23-26cm

S32175-3 Geophagus sp. cf. brasiliensis   JUVENIL
BAHIA - GEOPHAGUS
Bahia, East-Brazil, W, 22-25cm

South America Cichlids I   **51**

S32185-3 Geophagus sp. CAQUETA
CAQUETA - GEOPHAGUS
Rio Caqueta, Columbia, W, 18-20cm

S32190-3 Geophagus sp. CARONI
CARONI - GEOPHAGUS
Orinoco, Venezuela, W, 22-25cm

S32190-4 Geophagus sp. CARONI
CARONI - GEOPHAGUS
Orinoco, Venezuela, W, 22-25cm

S32195-2 Geophagus crassilabris   JUVENIL
PANAMA - GEOPHAGUS
Panama / Columbia, W, 21-23cm

S32195-4 Geophagus crassilabris   MALE
PANAMA - GEOPHAGUS
Panama / Columbia, W, 21-23cm

S32195-4 Geophagus crassilabris   FEMALE with Babies !!
PANAMA - GEOPHAGUS
Panama / Columbia, W, 21-23cm

S32200-3 Geophagus sp. CURUA
CURUA - GEOPHAGUS
Rio Curua, N/E-Brazil, W, 25-28cm

S32210-3 Geophagus grammepareius
CARONI - GEOPHAGUS
Caroni/Venezuela, W, 13-15cm

South America Cichlids I   © A.C.S. Glaser GmbH

S32225-4 Geophagus hondae-steindachneri
HUMPHEAD - GEOPHAGUS
Columbia, W, 18-22cm

S32235-4 Geophagus sp. INIRIDA
INIRIDA - GEOPHAGUS
Orinoco, Venezuela / Columbia, W, 23-25cm

S32250-3 Geophagus sp. MARABA (cf. altifrons)
MARABA - GEOPHAGUS
Tocantins, Brazil, W, 20-25cm

S32250-3 Geophagus sp. MARABA (cf. altifrons)
MARABA - GEOPHAGUS
Tocantins, Brazil, W, 20-25cm

S32260-3 Geophagus sp. MITU
MITU - GEOPHAGUS
Rio Vaupes, Columbia, W, 23-25cm

S32260-3 Geophagus sp. MITU
MITU - GEOPHAGUS
Rio Vaupes, Columbia, W, 23-25cm

S32275-4 Geophagus pellegrini
CHOCO - GEOPHAGUS
Choco, West-Columbia, W, 18-20cm

S32275-4 Geophagus pellegrini
CHOCO - GEOPHAGUS
Choco, West-Columbia, W, 18-20cm

South America **Cichlids I** 53

GEOPHAGUS PROXIMUS

1. Corydoras maculifer
2. Corydoras sp. axelrodi - deckeri

Diese und alle anderen demnächst in
These and all other coming soon in

Aqualog "all- corydoras"
reference fish of the world

South America **Cichlids I**   **55**

S32280-4 Geophagus sp. PINDARE
PINDARE GEOPHAGUS
Pindaiba-Area, Brazil, W, 15-17cm

S32285-2 Geophagus proximus JUVENIL
AMAZONAS - GEOPHAGUS
Amazonas / Santarem, W, 18-20cm

S32285-3 Geophagus proximus
AMAZONAS - GEOPHAGUS
Amazonas / Santarem, W, 18-20cm

S32285-4 Geophagus proximus
AMAZONAS - GEOPHAGUS
Amazonas / Santarem, W, 18-20cm

S32285-4 Geophagus proximus PAIR
AMAZONAS - GEOPHAGUS
Amazonas / Santarem, W, 18-20cm

S32295-2 Geophagus sp. "RED - PERU" JUVENIL
RED-PERU - GEOPHAGUS
Peru, W, 18-20cm

S32295-4 Geophagus sp. "RED - PERU" ADULT
RED-PERU - GEOPHAGUS
Peru, W, 18-20cm

S32305-3 Geophagus sp. RIO - MOJU
RIO-MOJU - GEOPHAGUS
Rio Moju, W, 14-16cm

South America **Cichlids I**

© A.C.S. Glaser GmbH

S32310-4 Geophagus sp. RIO - NEGRO II
RIO-NEGRO II - GEOPHAGUS
Rio Negro-System, North-Brazil, W, 18-20cm

S32320-4 Geophagus steindachneri (G. hondae) MALE
COLUMBIA - GEOPGHAGUS
Columbia, W, 20-23cm

S32320-4 Geophagus steindachneri (G. hondae)
COLUMBIA - GEOPGHAGUS   FEMALE with Babies !!
Columbia, W, 20-23cm

S32320-5 Geophagus steindachneri (G. hondae)
COLUMBIA - GEOPGHAGUS   GRANDFATHER
Columbia, W, 20-23cm

S32330-2 Geophagus surinamensis JUVENIL
SURINAM - GEOPHAGUS
Surinam, W, 18-20cm

S32330-3 Geophagus surinamensis
SURINAM - GEOPHAGUS
Surinam, W, 18-20cm

S32330-4 Geophagus surinamensis
SURINAM - GEOPHAGUS
Surinam, W, 18-20cm

S32345-4 Geophagus cf. surinamensis   "ovophil"
SURINAM - GEOPHAGUS nimmt die Eier ins Maul zum Ausbrüten !
Guayana,W,18-20cm,it take the Eggs into the Mouth for breeding !

South America **Cichlids I** 57

S32345-4 Geophagus cf. surinamensis "ovophil"
SURINAM - GEOPHAGUS nimmt die Eier ins Maul zum Ausbrüten !
Guayana,W,18-20cm,it take the Eggs into the Mouth for breeding !

S32335-4 Geophagus sp. surinamensis "larvophil"
SURINAM -GEOPHAGUS nimmt nur die Larven ins Maul bei Gefahr!
Guayana,W,15-17cm, it take only the larvs into the Mouth for care!

S32335-4 Geophagus sp. surinamensis "larvophil"
SURINAM - GEOPHAGUS nimmt nur die Larven ins Maul bei Gefahr!
Guayana,W,15-17cm, it take only the larvs into the Mouth for care!

S32345-5 Geophagus sp. surinam. "ovophil" FEMALE with Babies!
SURINAM - GEOPHAGUS nimmt die Eier ins Maul zum Ausbrüten !
Guayana,W,18-20cm,it take the Eggs into the Mouth for breeding !

S32360-4 Geophagus taeniopareius
VENEZUELA - GEOPHAGUS
Orinoco, Venezuela, W, 13-15cm

S32360-4 Geophagus taeniopareius PAIR
VENEZUELA - GEOPHAGUS
Orinoco, Venezuela, W, 13-15cm

S32365-4 Geophagus sp. TAPAJOS I "RED - CHEEK"
RED-CHEEK - GEOPHAGUS
Tapajos, Brazil, W, 15-17cm

S32365-4 Geophagus sp. TAPAJOS I "RED - CHEEK"
RED-CHEEK - GEOPHAGUS PAIR
Tapajos, Brazil, W, 15-17cm

South America Cichlids I   © A.C.S. Glaser GmbH

GEOPHAGUS SP. SURINAMENSIS

GEOPHAGUS SP. SURINAMENSIS

South America **Cichlids I** 59

S32370-4 Geophagus sp. TAPAJOS II
TAPAJOS II - GEOPHAGUS
Tapajos, Brazil, W, 15-17cm

S32375-4 Geophagus sp. TROMBETAS
TROMBETAS - GEOPHAGUS
Rio Trombetas, N/E-Brazil, W, 18-20cm

S32385-4 Geophagus sp. VENEZUELA
VENEZUELA - GEOPHAGUS
Venezuela, W, 22-25cm

S340004 Guianacara sp. CARONI
CARONI - GUIANACARA
Caroni, Venezuela, W, 13-15cm

S340004 Guianacara sp. CARONI
CARONI - GUIANACARA
Caroni, Venezuela, W, 13-15cm

S34020-3 Guianacara owroewefi
BLUE - GUIANACARA
Guayana, W, 13-15cm

S34025-3 Guianacara cf. owroewefi
SHOCO - GUIANACARA
Surinam, W, 12-14cm

S34025-3 Guianacara cf. owroewefi
SHOCO - GUIANACARA
Surinam, W, 12-14cm

60 South America Cichlids I © A.C.S. Glaser GmbH

S34035-4 Guianacara sp. "RED - CHEEK" ADULT
RED-CHEEK - GUIANACARA
Fr.-Guayana, W, 13-15cm

S34035-4 Guianacara sp. "RED - CHEEK" ADULT
RED-CHEEK - GUIANACARA
Fr.-Guayana, W, 13-15cm

S34035-4 Guianacara sp. "RED - CHEEK" ADULT / PAIR
RED-CHEEK - GUIANACARA
Fr.-Guayana, W, 13-15cm

S34205-4 Gymnogeophagus sp. "BAHIA - RED" FEMALE
BAHIA - GEOPHAGUS
South - Brazil, W, 18-22cm

S34205-4 Gymnogeophagus sp. "BAHIA - RED"
BAHIA - GEOPHAGUS PAIR with Eggs
South - Brazil, W, 18-22cm

S34205-4 Gymnogeophagus sp. "BAHIA - RED"
BAHIA - GEOPHAGUS PAIR with Babies!
South - Brazil, W, 18-22cm

S34210-2 Gymnogeophagus balzani JUVENIL
BALZANI - CICHLID
Paraguay / Argentina, W, 15-17cm

S34210-3 Gymnogeophagus balzani
BALZANI - CICHLID
Paraguay / Argentina, W, 15-17cm

South America **Cichlids I**

S34210-4 Gymnogeophagus balzani   FEMALE with Babies !
BALZANI - CICHLID
Paraguay / Argentina, W, 15-17cm

S34210-5 Gymnogeophagus balzani   ADULT - MALE
BALZANI - CICHLID
Paraguay / Argentina, W, 15-17cm

S34215-4 Gymnogeophagus sp. BRAZIL I   Male
BRAZIL I - GEOPHAGUS
South - Brazil, W, 13-15cm

S34220-3 Gymnogeophagus sp. BRAZIL II
BRAZIL II - GEOPHAGUS
South - Brazil, W, 8-10cm

S34230-3 Gymnogeophagus gymnogenys   MALE
REDFIN - GEOPHAGUS
South - Brazil, W, 16-18cm

S34230-3 Gymnogeophagus gymnogenys   FEMALE
REDFIN - GEOPHAGUS
South - Brazil, W, 16-18cm

S34230-5 Gymnogeophagus gymnogenys
REDFIN - GEOPHAGUS
South - Brazil, W, 16-18cm

S34240-3 Gymnogeophagus labiatus   MALE
LABIATUS - GEOPHAGUS
South - Brazil, W, 16-18cm

South America   Cichlids I   © A.C.S. Glaser GmbH

S34240-3 Gymnogeophagus labiatus FEMALE
LABIATUS - GEOPHAGUS
South - Brazil, W, 16-18cm

S34245-4 Gymnogeophagus sp. labiatus MALE
ROUNDHEAD - GEOPHAGUS
South - Brazil, W, 16-18cm

S34245-4 Gymnogeophagus sp. labiatus PAIR
ROUNDHEAD - GEOPHAGUS
South - Brazil, W, 16-18cm

S34255-2 Gymnogeophagus meridionalis JUVENIL
MERIDIO - GEOPHAGUS
South - Brazil, W, 13-15cm

S34255-3 Gymnogeophagus meridionalis
MERIDIO - GEOPHAGUS
South - Brazil, W, 13-15cm

S34255-4 Gymnogeophagus meridionalis
MERIDIO - GEOPHAGUS
South - Brazil, W, 13-15cm

S34255-4 Gymnogeophagus meridionalis PAIR
MERIDIO - GEOPHAGUS   bei der Brutpflege !
South - Brazil, W, 13-15cm   by breeding care !

S34265-4 Gymnogeophagus rhabdotus
RHABDOTUS - GEOPHAGUS
Brazil, W, 14-16cm

South America **Cichlids I** 63

GYMNOGEOPHAGUS BALZANI   MALE

GYMNOGEOPHAGUS GYMNOGENYS   MALE

South America **Cichlids I**

S34275-3 Gymnogeophagus setequedas   MALE
QUEDAS - GEOPHAGUS
South - Brazil, W, 12-13cm

S34275-3 Gymnogeophagus setequedas   FEMALE
QUEDAS - GEOPHAGUS
South - Brazil, W, 12-13cm

S78205-3 Retroculus lapidifer
MARBLE - RETROCULUS
Tocantins, Araguaia, Brazil, W, 23-25cm

S78205-3 Retroculus lapidifer   PAIR
MARBLE - RETROCULUS
Tocantins, Araguaia, Brazil, W, 23-25cm

S78205-4 Retroculus lapidifer
MARBLE - RETROCULUS
Tocantins, Araguaia, Brazil, W, 23-25cm

S78215-2 Retroculus xinguensis   JUVENIL
XINGU - RETROCULUS
Xingu, Brazil, W, 23-25cm

S78215-4 Retroculus xinguensis
XINGU - RETROCULUS
Xingu, Brazil, W, 23-25cm

S78215-4 Retroculus xinguensis
XINGU - RETROCULUS
Xingu, Brazil, W, 23-25cm

South America **Cichlids I**

© A.C.S. Glaser GmbH

S85305-2 Satanoperca acuticeps   JUVENIL
FOUR-SPOT - GEOPHAGUS
Santarem, Amazonas, W, 23-25cm

S85305-3 Satanoperca acuticeps
FOUR-SPOT - GEOPHAGUS
Santarem, Amazonas, W, 23-25cm

S85315-4 Satanoperca sp. COLUMBIA
COLUMBIA - SATANOPERCA
Rio Vaupes, Columbia, W, 23-25cm

S85325-4 Satanoperca daemon
THREE - SPOT - GEOPHAGUS
Orinoco-Area, W, 23-25cm

S85325-4 Satanoperca daemon   Schreckfärbung
THREE - SPOT - GEOPHAGUS   fright - color
Orinoco-Area, W, 23-25cm

S85325-4 Satanoperca daemon   Aggressions-Färbung
THREE - SPOT - GEOPHAGUS   aggressive - color
Orinoco-Area, W, 23-25cm

S85335-4 Satanoperca sp. FRENCH-GUAYANA
FRENCH-GUAYANA - SATANOPERCA
Fr.-Guayana, Cirique St. Anne, W, 24-26cm

S85350-2 Satanoperca jurupari
EARTH - EATER
Santarem, Brazil, W, 22-25cm

South America **Cichlids I**

S85350-3 Satanoperca jurupari
EARTH - EATER
Santarem, Brazil, W, 22-25cm

S85350-4 Satanoperca jurupari
EARTH - EATER
Santarem, Brazil, W, 22-25cm

S85355-2 Satanoperca sp. jurupari
EARTH - EATER   gefangen an der Grenze von
W, 20-22cm   keeped it at border Fr.Guay.+Surinam

S85355-4 Satanoperca sp. jurupari
EARTH - EATER   gefangen an der Grenze von
W, 20-22cm   keeped it at border Fr.Guay.+Surinam

S85365-4 Satanoperca leucosticta
CUYUNI - SATANOPERCA
Cuyuari, Venezuela, W, 23-25cm

S85365-5 Satanoperca leucosticta
CUYUNI - SATANOPERCA
Cuyuari, Venezuela, W, 23-25cm

S85365-4 Satanoperca sp. leucosticta
META - SATANOPERCA
Rio Meta, Columbia, W, 23-25cm

S85365-4 Satanoperca sp. leucosticta   Pair
META - SATANOPERCA
Rio Meta, Columbia, W, 23-25cm

SATANOPERCA DAEMON "ADULT"

South America **Cichlids I**

1. **Hier ist ihre Heimat, fotografiert während der Trockenzeit. In der Regenzeit ist der Wasserstand 1-2 m höher.**

   This is its native environment, photographed during the dry season. During the rainy season the water level is 1 - 2 metres higher.

2. **In großen Freilandanlagen, wie hier in Brasilien, werden sie für den Export gezüchtet.**

   They are bred for export in large outdoor farms, such as here in Brazil.

S85370-4 Satanoperca cf. leucosticta
PEARL - SATANOPERCA
Brazil (?) unbekannte Herkunft / unknown place 20cm

S85370-4 Satanoperca cf. leucosticta
PEARL - SATANOPERCA
Brazil (?) unbekannte Herkunft / unknown place 20cm

S85375-3 Satanoperca lilith
ONE-SPOT - GEOPHAGUS
Rio Negro, Brazil, W, 23-25cm

S85375-3 Satanoperca lilith   Schreckfärbung / fright-color
ONE-SPOT - GEOPHAGUS
Rio Negro, Brazil, W, 23-25cm

S85385-4 Satanoperca mapiritensis
ORINOCO - SATANOPERCA
Orinoco, Venezuela, W, 23-25cm

S85385-4 Satanoperca mapiritensis
ORINOCO - SATANOPERCA
Orinoco, Venezuela, W, 23-25cm

S85395-4 Satanoperca pappaterra
PANTANAL - SATANOPERCA
Pantanal, Brazil, W, 18-20cm

S85395-4 Satanoperca pappaterra   PAIR
PANTANAL - SATANOPERCA   Nachtfärbung
Pantanal, Brazil, W, 18-20cm   color at the night

South America **Cichlids I**

S85405-3 Satanoperca sp. "REDLIP"   Gruppe / Group
REDLIP - SATANOPERCA
Fr.-Guayana, W, 18-20cm

S85405-4 Satanoperca sp. "REDLIP"
REDLIP - SATANOPERCA
Fr.-Guayana, W, 18-20cm

S85405-5 Satanoperca sp. "REDLIP"
REDLIP - SATANOPERCA
Fr.-Guayana, W, 18-20cm

S85410-4 Satanoperca sp. RIO - XINGU (lilith ??)
XINGU - SATANOPERCA
Rio Xingu, Brazil, W, 16-18cm

S85420-4 Satanoperca sp. UAUPES
UAUPES - SATANOPERCA
Rio Tiquie, W, 16-18cm

S85415-4 Satanoperca sp. GUAYANA
GUAYANA - SATANOPERCA
Guayana, W, 14-16cm

S85415-4 Satanoperca sp. GUAYANA
GUAYANA - SATANOPERCA
Guayana, W, 14-16cm

S93004-3 Teleocichla sp. PARU
PARU - TELEO
Rio Paru do Oeste; North-Brazil, W, 12-13cm

South America Cichlids I   © A.C.S. Glaser GmbH

S93005-3 Teleocichla sp. I   MALE
I - TELEO
Rio Xingu, Brazil, W, 5-6cm

S93005-3 Teleocichla sp. I   FEMALE
I - TELEO
Rio Xingu, Brazil, W, 4-5cm

S93010-3 Teleocichla sp. II "BLACK"
BLACK - TELEO II
Rio Xingu, Brazil, W, 13-15cm

S93010-3 Teleocichla sp. II "BLACK"
BLACK - TELEO II
Rio Xingu, Brazil, W, 13-15cm

S93015-3 Teleocichla sp. III "DOT"
DOT - TELEO III
Rio Xingu, Brazil, W, 5-6cm

S93015-3 Teleocichla sp. III "DOT"
DOT - TELEO III
Rio Xingu, Brazil, W, 5-6cm

S93020-3 Teleocichla sp. IV "GOBIO"   MALE
GOBIO - TELEO IV
Rio Xingu, Brazil, W, 7-8cm

S93020-3 Teleocichla sp. IV "GOBIO"   FEMALE
GOBIO - TELEO IV
Rio Xingu, Brazil, W, 5-6cm

South America **Cichlids I**

S93030-3 Teleocichla centrarchus MALE
CENTRO - TELEO
Rio Xingu, N/E-Brazil, W, 10-12cm

S93030-3 Teleocichla centrarchus FEMALE
CENTRO - TELEO
Rio Xingu, N/E-Brazil, W, 8-9cm

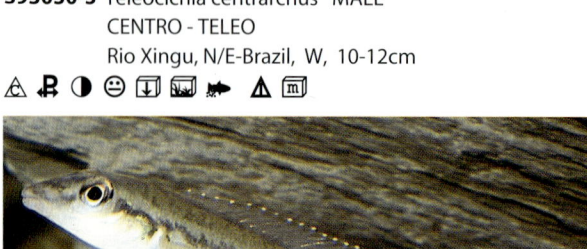

S93035-4 Teleocichla cinderella MALE
CINDI - TELEO
Rio Tocantins, N/E-Brazil, W, 10-12cm

S93035-4 Teleocichla cinderella FEMALE
CINDI - TELEO
Rio Tocantins, N/E-Brazil, W, 8-9cm

S93040-3 Teleocichla cf. cinderella
BROADBAND - TELEO
Brazil, W, 8-11cm

S93050-2 Teleocichla gephyrogramma JUVENIL
GEPHYRO - TELEO
Rio Xingu, N/E-Brazil, W, M.7-8cm + F.5-6cm

S93050-3 Teleocichla gephyrogramma MALE
GEPHYRO - TELEO
Rio Xingu, N/E-Brazil, W, M.7-8cm + F.5-6cm

S93050-3 Teleocichla gephyrogramma FEMALE
GEPHYRO - TELEO
Rio Xingu, N/E-Brazil, W, M.7-8cm + F.5-6cm

South America Cichlids I © A.C.S. Glaser GmbH

S93060-3 Teleocichla monogramma MALE
MONOGRAMMA - TELEO
Rio Xingu, N/E-Brazil, W, 10-12cm

S93060-3 Teleocichla monogramma FEMALE
MONOGRAMMA - TELEO
Rio Xingu, N/E-Brazil, W, 8-9cm

S93063-3 Teleocichla sp. monogramma FEMALE
MONOGRAMMA - TELEO
Rio Xingu, N/E-Brazil, W, 6-7cm

S93070-4 Teleocichla prionogenys MALE
PRIONO - TELEO
Rio Tapajos, N/E-Brazil, W, 11-13cm

S93070-4 Teleocichla prionogenys FEMALE
PRIONO - TELEO
Rio Tapajos, N/E-Brazil, W, 7-8cm

S93075-4 Teleocichla proselytus MALE
LYTO - TELEO
Rio Tapajos, N/E-Brazil, W, 9-11cm

S93075-4 Teleocichla proselytus FEMALE
LYTO - TELEO
Rio Tapajos, N/E-Brazil, W, 7-8cm

S93075-4 Teleocichla proselytus FEMALE bei Brutpflege
LYTO - TELEO   by breeding care
Rio Tapajos, N/E-Brazil, W, 7-8cm

South America **Cichlids I**

S93085-3 Teleocichla sp.
SPOTLINE - TELEO
Brazil, W, 10-12cm

S93085-3 Teleocichla sp.
SPOTLINE - TELEO
Brazil, W, 10-12cm

S93095-4 Teleocichla sp. XINGU III   MALE
XINGU III - TELEO
Rio Xingu, N/E-Brazil, W, 7-8cm

S93095-4 Teleocichla sp. XINGU III   FEMALE
XINGU III - TELEO
Rio Xingu, N/E-Brazil, W, 5-6cm

S93100-3 Teleocichla sp. XINGU IV   MALE
XINGU IV - TELEO
Rio Xingu, N/E-Brazil, W, 7-8cm

S93100-3 Teleocichla sp. XINGU IV   FEMALE
XINGU IV - TELEO
Rio Xingu, N/E-Brazil, W, 5-6cm

S93090-4 Teleocichla sp. XINGU II   MALE
XINGU II - TELEO
Rio Xingu, N/E-Brazil, W, 14-16cm

S93090-4 Teleocichla sp. XINGU II   FEMALE
XINGU II - TELEO
Rio Xingu, N/E-Brazil, W, 10-11cm

South America **Cichlids I**

© A.C.S. Glaser GmbH

S99755-1 Uaru amphiacanthoides   BABY = whitespot - color
WEDGESPOT - CICHLID
all Amazonas - Area, W, 23-25cm

S99755-2 Uaru amphiacanthoides   JUVENIL
WEDGESPOT - CICHLID
all Amazonas - Area, W, 23-25cm

S99755-4 Uaru amphiacanthoides
WEDGESPOT - CICHLID
all Amazonas - Area, W, 23-25cm

S99755-4 Uaru amphiacanthoides   MALE
WEDGESPOT - CICHLID
all Amazonas - Area, W, 23-25cm

S99755-4 Uaru amphiacanthoides   FEMALE
WEDGESPOT - CICHLID
all Amazonas - Area, W, 23-25cm

S99755-4 Uaru amphiacanthoides   PAIR
WEDGESPOT - CICHLID
all Amazonas - Area, W, 23-25cm

S99755-5 Uaru amphiacanthoides
WEDGESPOT - CICHLID
all Amazonas - Area, W, 23-25cm

Cichlasoma (Heros) festae blue   BLUE FESTAE - CICHLID
alle anderen demnächst in / all other coming soon in
" SOUTHAMERICAN CICHLIDS III"

South America Cichlids I

S00103-3 Acarichthys geayi (new name: Guianacara geayi)
GEAYI - ACARICHTHYS
Brazil / Columbia, W, 16-18cm

S00106-2 Acarichthys heckelii JUVENIL
HECKEL - ACARICHTHYS
Amazonas (Santarem), W, 20-22cm

S00106-3 Acarichthys heckelii
HECKEL - ACARICHTHYS
Amazonas (Santarem), W, 20-22cm

S00106-4 Acarichthys heckelii PAIR
HECKEL - ACARICHTHYS
Amazonas (Santarem), W, 20-22cm

S06258-3 Astronotus ocellatus WILD-FORM
OSCAR WILD-FORM
Southamerica, 30-35cm

S06267-4 Astronotus ocellatus ALBINO
ALBINO - OSCAR
Southamerica, Z, 30-35cm

S06268-4 Astronotus ocellatus WILD-FORM - BIG-SPOT
OSCAR WILD-FORM
Southamerica, 30-35cm

S06261-4 Astronotus sp. ocellatus "MARBLE"
MARBLE - OSCAR
Southamerica, 30-35cm

South America **Cichlids I**

© A.C.S. Glaser GmbH

S06265-4 Astronotus sp. ocellatus "RED - PEARL"
RED-PEARL - OSCAR
Southamerica, 20-25cm

S06269-4 Astronotus sp. ocellatus plane-colour (not red)
BLACKFIN - OSCAR
Southamerica, 20-25cm

S06269-4 Astronotus sp. ocellatus plane-colour (not red)
BLACKFIN - OSCAR   PAIR
Southamerica, 20-25cm

S07703-4 Biotodoma cupido
CUPIDO - CICHLID
Tocantins, Brazil, W, 11-13cm

S07705-4 Biotodoma cf. cupido
BELEM - CUPIDO
Tocantins, Brazil, W, 11-13cm

S07705-4 Biotodoma cf. cupido
BELEM - CUPIDO
Tocantins, Brazil, W, 11-13cm

S07715-3 Biotodoma wavrini
WAVRINI - BIOTODOMA
Columbia, W, 12-14cm

S07717-3 Biotodoma cf. wavrini
ORINOCO - BIOTODOMA
Orinoco - Area, W, 13-15cm

South America Cichlids I

S07717-3 Biotodoma cf. wavrini
ORINOCO - BIOTODOMA
Orinoco - Area, W, 13-15cm

S11908-3 Chaetobranchopsis bitaeniatus
TWO-SPOT - CHAETOBRANCHOPSIS
Paraguay / Argentina, W, 8-10cm

S11908-4 Chaetobranchopsis bitaeniatus
TWO-SPOT - CHAETOBRANCHOPSIS
Paraguay / Argentina, W, 8-10cm

S11915-4 Chaetobranchopsis flavescens
FLAVI - CHAETOBRANCHOPSIS
Paraguay, W, 20-25cm

S11915-4 Chaetobranchopsis flavescens   PAIR
FLAVI - CHAETOBRANCHOPSIS
Paraguay, W, 20-25cm

S11925-4 Chaetobranchopsis orbicularis
ORBI - CHAETOBRANCHOPSIS
Amazonas - Area, W, 11-13cm

S11927-4 Chaetobranchopsis sp. orbicularis
BLACK-BAND - CHAETOBRANCHOPSIS
Bolivia, W, 12-15cm

S11935-4 Chaetobranchopsis spectabilis
SPECTA - CHAETOBRANCHOPSIS
Venezuela, W, 14-16cm

South America Cichlids I   © A.C.S. Glaser GmbH

S32373-3 Geophagus sp. "TOCANTINS" MALE
TOCANTINS - GEOPHAGUS
0019/ 81-1 Brazil: Rio Tocantins/Araguaia, W, 17cm
Foto: Uwe Werner

S32373-3 Geophagus sp. "TOCANTINS" FEMALE
TOCANTINS - GEOPHAGUS
0020/ 81-2 Brazil: Rio Tocantins/Araguaia, W, 17cm
Foto: Uwe Werner

S32373-4 Geophagus sp. "TOCANTINS" FEMALE with babies !
TOCANTINS - GEOPHAGUS
0021/ 81-3 Brazil: Rio Tocantins/Araguaia, W, 17cm
Foto: Uwe Werner

S13748-2 Cichla sp. "PERU" JUVENIL
PERU - CICHLA
0022/ 81-4 Peru, (other descriptions impossible still yet)
Foto: Uwe Werner

S13710-3 Cichla monoculus
AMAZONIA - CICHLA
0023/ 81-5 Amazonas-area, Brazil, Peru, Bolivia, W, 45cm
Foto: Uwe Werner

S23710-3 Crenicichla phaiospilus FEMALE
SPILUS - CRENICICHLA
0024/ 81-6 Rio Xingu, north-east-Brazil, W, 30cm
Foto: Frank Warzel

S23835-3 Crenicichla tigrina MALE
TIGRINA - CRENICICHLA
0025/ 81-7 Rio Trombetas, north-east-Brazil, W, 35cm
Foto: Frank Warzel

S06264-4 Astronotus ocellatus "RED"
RED - OSCAR
0026/ 81-8 Breeding-form, Southamerica, Z, 30-35cm
Foto: Erwin Schraml

South America Cichlids I   81

S13701-4 Cichla cf. nigrolineata
0002/82-1 W, 50cm
Orinoco, Venezuela + Columbia
Foto: Uwe Werner

S32371-4 Geophagus sp. „Tapajos III", Male
0056/82-2 Tapajos-Tränenstrich/Tearstripe Eartheater
Brazil, Rio Tapajos
Foto: H. Morche

S32274-4 Geophagus sp. „Paru", Male
0057/82-4 Rio-Paru-Eartheater
North Brazil, Rio Paru, W, 12 cm
Foto: H. Morche

82 South America Cichlids I © A.C.S. Glaser GmbH

In regelmäßigen Abständen erscheinen Ergänzungen mit neuen Fisch-Bildern, die können Sie

hier

einkleben, damit Sie immer "up-to-date" sind.

Supplements featuring new fish-photographs will be issued on regular basis. Stick them in

here

so that your collection is always up to date.

In regelmäßigen Abständen erscheinen Ergänzungen mit neuen Fisch-Bildern, die können Sie

hier

einkleben, damit Sie immer "up-to-date" sind.

Supplements featuring new fish-photographs will be issued on regular basis. Stick them in

here

so that your collection is always up to date.

© A.C.S. Glaser GmbH

In regelmäßigen Abständen erscheinen Ergänzungen mit neuen Fisch-Bildern, die können Sie

hier

einkleben, damit Sie immer "up-to-date" sind.

Supplements featuring new fish-photographs will be issued on regular basis. Stick them in

here

so that your collection is always up to date.

South America **Cichlids I** © **A.C.S. Glaser GmbH**

In regelmäßigen Abständen erscheinen Ergänzungen mit neuen Fisch-Bildern, die können Sie

hier

einkleben, damit Sie immer "up-to-date" sind.

Supplements featuring new fish-photographs will be issued on regular basis. Stick them in

here

so that your collection is always up to date.

© A.C.S. Glaser GmbH

In regelmäßigen Abständen erscheinen Ergänzungen mit neuen Fisch-Bildern, die können Sie

hier

einkleben, damit Sie immer "up-to-date" sind.

Supplements featuring new fish-photographs will be issued on regular basis. Stick them in

here

so that your collection is always up to date.

South America **Cichlids I** © A.C.S. Glaser GmbH

In regelmäßigen Abständen erscheinen Ergänzungen mit neuen Fisch-Bildern, die können Sie

hier

einkleben, damit Sie immer "up-to-date" sind.

Supplements featuring new fish-photographs will be issued on regular basis. Stick them in

here

so that your collection is always up to date.

© A.C.S. Glaser GmbH

In regelmäßigen Abständen erscheinen Ergänzungen mit neuen Fisch-Bildern, die können Sie

hier

einkleben, damit Sie immer "up-to-date" sind.

Supplements featuring new fish-photographs will be issued on regular basis. Stick them in

here

so that your collection is always up to date.

South America **Cichlids I**　　　© **A.C.S. Glaser GmbH**

In regelmäßigen Abständen erscheinen Ergänzungen mit neuen Fisch-Bildern, die können Sie

hier

einkleben, damit Sie immer "up-to-date" sind.

Supplements featuring new fish-photographs will be issued on regular basis. Stick them in

here

so that your collection is always up to date.

© A.C.S. Glaser GmbH

In regelmäßigen Abständen erscheinen Ergänzungen mit neuen Fisch-Bildern, die können Sie

hier

einkleben, damit Sie immer "up-to-date" sind.

Supplements featuring new fish-photographs will be issued on regular basis. Stick them in

here

so that your collection is always up to date.

In regelmäßigen Abständen erscheinen Ergänzungen mit neuen Fisch-Bildern, die können Sie

hier

einkleben, damit Sie immer "up-to-date" sind.

Supplements featuring new fish-photographs will be issued on regular basis. Stick them in

here

so that your collection is always up to date.

© A.C.S. Glaser GmbH

In regelmäßigen Abständen erscheinen Ergänzungen mit neuen Fisch-Bildern, die können Sie

hier

einkleben, damit Sie immer "up-to-date" sind.

Supplements featuring new fish-photographs will be issued on regular basis. Stick them in

here

so that your collection is always up to date.

South America **Cichlids I** © A.C.S. Glaser GmbH

In regelmäßigen Abständen erscheinen Ergänzungen mit neuen Fisch-Bildern, die können Sie

hier

einkleben, damit Sie immer "up-to-date" sind.

Supplements featuring new fish-photographs will be issued on regular basis. Stick them in

here

so that your collection is always up to date.

© A.C.S. Glaser GmbH

In regelmäßigen Abständen erscheinen Ergänzungen mit neuen Fisch-Bildern, die können Sie

hier

einkleben, damit Sie immer "up-to-date" sind.

Supplements featuring new fish-photographs will be issued on regular basis. Stick them in

here

so that your collection is always up to date.

South America **Cichlids I**

© A.C.S. Glaser GmbH

In regelmäßigen Abständen erscheinen Ergänzungen mit neuen Fisch-Bildern, die können Sie

hier

einkleben, damit Sie immer "up-to-date" sind.

Supplements featuring new fish-photographs will be issued on regular basis. Stick them in

here

so that your collection is always up to date.

© A.C.S. Glaser GmbH

In regelmäßigen Abständen erscheinen Ergänzungen mit neuen Fisch-Bildern, die können Sie

hier

einkleben, damit Sie immer "up-to-date" sind.

Supplements featuring new fish-photographs will be issued on regular basis. Stick them in

here

so that your collection is always up to date.

South America **Cichlids I** © A.C.S. Glaser GmbH

In regelmäßigen Abständen erscheinen Ergänzungen mit neuen Fisch-Bildern, die können Sie

hier

einkleben, damit Sie immer "up-to-date" sind.

Supplements featuring new fish-photographs will be issued on regular basis. Stick them in

here

so that your collection is always up to date.

In regelmäßigen Abständen erscheinen Ergänzungen mit neuen Fisch-Bildern, die können Sie

Supplements featuring new fish-photographs will be issued on regular basis. Stick them in

© A.C.S. Glaser GmbH

INDEX
Code - numbers

S00103	Acarichthys geayi (Guianacara geyai)	GEAYI-ACARICHTHYS	78
S00106	Acarichthys heckelii	HECKEL-ACARICHTHS	78
S06258	Astronotus ocellatus	OSCAR-WILDFORM	78
S06261	Astronotus ocellatus sp. "MARBLE"	MARBLE-OSCAR	78
S06265	Astronotus ocellatus sp. "RED-PEARL"	RED-PEARL-OSCAR	79
S06267	Astronotus ocellatus "ALBINO"	ALBINO-OSCAR	78
S06268	Astronotus ocellatus WILDFORM-BIGSPOT	OSKAR WILD-FORM	78
S06269	Astronotus ocellatus sp. "PLANE-COLOUR"	BLACKFIN-OSCAR	79
S07703	Biotodoma cupido	CUPIDO-CICHLID	79
S07705	Biotodoma cupido cf.	BELEM-BIOTODOMA	79
S07715	Biotodoma wavrini	WAVRINI-BIOTODOMA	79
S07717	Biotodoma wavrini cf.	ORINOCO-BIOTODOMA	79,80
S11908	Chaetobranchopsis bitaeniatus	TWO-SPOT-CHAETOBRANCHOPSIS	80
S11915	Chaetobranchopsis flavescens	FLAVI-CHAETOBRANCHOPSIS	80
S11925	Chaetobranchopsis orbicularis	ORBI-CHAETOBRANCHOPSIS	80
S11927	Chaetobranchopsis orbicularis sp.	BLACK-BAND-CHAETOBRANCHOPSIS	80
S11935	Chaetobranchopsis spectabilis	SPECTA-CHAETOBRANCHOPSIS	80
S13703	Cichla sp. "ARAGUAIA" III	ARAGUAIA-CICHLA	7
S13718	Cichla ocellaris	PIKE-CICHLID	7,12
S13725	Cichla ocellaris cf. "RIO PRETO"	PRETO-PIKE-CICHLID	7
S13735	Cichla orinocensis	ORINOCO-PIKE-CICHLID	7
S13745	Cichla orinocensis cf. "RIO-INIRIDA"	INIRIDA-PIKE-CICHLID	8
S13748	Cichla orinocensis cf. "RIO-VAUPES"	VAUPES-PIKE-CICHLID	8
S13755	Cichla sp. "TAPAJOS" I	TAPAJOS-CICHLA	8
S13758	Cichla temensis	TEMENSIS-PIKE-CICHLID	8
S13760	Cichla temensis cf. "CARONI"	CARONI-CICHLA	8
S13762	Cichla temensis cf. "ARAGUAIA"	ARAGUAIA-CICHLA	9
S13763	Cichla temensis cf. "CAURA"	CAURA-CICHLA	9
S13765	Cichla temensis cf. "VAUPES"	VAUPES-CICHLA	9
S13770	Cichla sp. "XINGU" II	XINGU II-CICHLA	9
S23310	Crenicichla acutirostris	ACUTIS-CRENICICHLA	9
S23315	Crenicichla albopunctata	SPOTTED-CRENICICHLA	10
S23320	Crenicichla sp. "ALTA"	ALTA-CRENICICHLA	10
S23325	Crenicichla anthurus	ANTHURUS-CRENICICHLA	10
S23330	Crenicichla sp. "APPROUAGE"	APPROUAGE-CRENICICHLA	10
S23335	Crenicichla sp. "ARAPIUNS"	ARAPIUNS-CRENICICHLA	11
S23340	Crenicichla sp. "ATABAPO"	ATABAPO-CRENICICHLA	11
S23350	Crenicichla sp. "BELEM"	BELEM-PIKE-CICHLID	11
S23355	Crenicichla sp. "BELLY-CRAWLER"	BELLY-CRAWLER-CRENICICHLA	11
S23360	Crenicichla sp. "BOCON"	BOCON-CRENICICHLA	11
S23370	Crenicichla cametana	TOCANTINS-CRENICICHLA	14
S23375	Crenicichla cardiostigma	RIO-BRANCO-CRENICICHLA	14
S23378	Crenicichla cardiostigma (pterogramma?)	PTEROGRAMMA-CRENICICHLA	14
S23380	Crenicichla sp. "CASIQUIARE"	CASIQUIARE-CRENICICHLA	14
S23385	Crenicichla cincta	CINCTA-CRENICICHLA	14
S23390	Crenicichla compressiceps	COMPRESS-CRENICICHLA	15,21
S23395	Crenicichla cyanonotus	CYANO-CRENICICHLA	15
S23400	Crenicichla cyclostoma	CYCLOS-CRENICICHLA	15
S23415	Crenicichla dorsocellata	CAMPOS-CRENICICHLA	15
S23425	Crenicichla edithae	EDITH-CRENICICHLA	16
S23435	Crenicichla frenata	TRINIDAD-CRENICICHLA	16
S23440	Crenicichla frenata var.	FRENATA-CRENICICHLA	16
S23450	Crenicichla geayi	GEAYI-CRENICICHLA	16
S23455	Crenicichla geayi cf.	BRIGHT-BLUE-CRENICICHLA	17
S23465	Crenicichla sp. "GUANIAMO"	GUANIAMO-CRENICICHLA	17
S23470	Crenicichla sp. "GUARIQUITO"	GUARIQUITO-CRENICICHLA	17
S23475	Crenicichla sp. "GUARIQUITO" var.	AYACUCHO-CRENICICHLA	17,18
S23480	Crenicichla sp. "GUAYANA"	DEMERARA-CRENICICHLA	18
S23490	Crenicichla heckeli	HECKEL-CRENICICHLA	18
S23500	Crenicichla sp. "INIRIDA" I	INIRIDA I-CRENICICHLA	18
S23505	Crenicichla sp. "INIRIDA" II	INIRIDA II-CRENICICHLA	18
S23510	Crenicichla sp. "INIRIDA" III "RED-SPOT"	INIRIDA III-"RED-SPOT"-Crenicichla	19

INDEX
Code - numbers

Code	Species	Common Name	Page
S23520	Crenicichla sp. "JABUTI"	JABUTI-CRENICICHLA	19
S23525	Crenicichla jaguarensis	JAGUAR-CRENICICHLA	19
S23530	Crenicichla jegui	JEGUI-CRENICICHLA	19
S23535	Crenicichla jegui sp.	NORTH-EAST-CRENICICHLA	20
S23540	Crenicichla johanna	JOHANNA-CRENICICHLA	20
S23545	Crenicichla johanna var. cf.	JOHANNA-VAR.-CRENICICHLA	20
S23560	Crenicichla labrina	LABRINA-CRENICICHLA	20
S23565	Crenicichla labrina sp.	AMAZONAS-CRENICICHLA	20
S23570	Crenicichla lenticulata	LENTICULATA-CRENICICHLA	22
S23575	Crenicichla lepidota	LEPIDOTA-CRENICICHLA	22
S23580	Crenicichla lepidota "REAL"	REAL-LEPIDOTA-CRENICICHLA	22
S23590	Crenicichla lucius	LUCIUS-CRENICICHLA	22,23
S23595	Crenicichla lugubris	LUGU-CRENICICHLA	23,24
S23596	Crenicichla lugubris var.	RED-MOUTH-CRENICICHLA	23
S23600	Crenicichla lugubris cf. "ATABAPO"	ATABAPO-CRENICICHLA	23
S23610	Crenicichla macrophthalma	MACRO-CRENICICHLA	26
S23615	Crenicichla sp. "MAICURU"	MAICURU-CRENICICHLA	26
S23620	Crenicichla sp. "MANAUS"	MANAUS-CRENICICHLA	26
S23625	Crenicichla marmorata	MARBLE-CRENICICHLA	26,27
S23630	Crenicichla sp. "MARMORATA"	MARMORATA-CRENICICHLA	27
S23635	Crenicichla sp. "MATO-GROSSO"	MATO-GROSSO-CRENICICHLA	27
S23640	Crenicichla menezesi sp.	MENEZESI-CRENICICHLA	27,28
S23645	Crenicichla multispinosa	MULTI-CRENICICHLA	28
S23655	Crenicichla nothophthalmus sp.	NEGRO-CRENICICHLA	28
S23660	Crenicichla notophthalmus	NORTH-CRENICICHLA	25,28
S23663	Crenicichla notophthalmus cf.	BIG-POINT-CRENICICHLA	28
S23664	Crenicichla nothophthalmus sp. "UAUPES"	UAUPES-CRENICICHLA	31
S23670	Crenicichla sp. "ORANGE"-Xingu/Belem	ORANGE-XINGU-CRENICICHLA	31
S23675	Crenicichla "ORINOCO"	ORINOCO-CRENICICHLA	31
S23680	Crenicichla sp. "ORINOCO"	MANY-SPOT-CRENICICHLA	31
S23690	Crenicichla sp. "PACAYA"	PACAYA-CRENICICHLA	31
S23695	Crenicichla percna	PERCNA-CRENICICHLA	32
S23705	Crenicichla sp. "PERNAMBUCO"	PERNAMBUCO-CRENICICHLA	32
S23710	Crenicichla phaiospilus	SPILUS-CRENICICHLA	30,32
S23715	Crenicichla sp. "PINDARE"	PINDARE-CRENICICHLA	33
S23720	Crenicichla proteus	PERU-CRENICICHLA	33
S23725	Crenicichla punctata	PUNCTATA-CRENICICHLA	33
S23735	Crenicichla sp. "RED-BELLY"	RED-BELLY-Crenicichla	33
S23737	Crenicichla regani "DAS MORTES"	MORTES-CRENICICHLA	34,35
S23738	Crenicichla regani "RIO-GUAMA"	GUAMA-CRENICICHLA	35
S23739	Crenicichla regani "RIO-ACARA"	ACARA-CRENICICHLA	35
S23740	Crenicichla regani	REGANI-CRENICICHLA	35
S23741	Crenicichla regani "RIO-NEGRO"	RIO-NEGRO-CRENICICHLA	34,36
S23742	Crenicichla regani "TEFE"	TEFE-CRENICICHLA	36
S23743	Crenicichla regani "TROMBETAS"	TROMBETAS-CRENICICHLA	36
S23744	Crenicichla regani "GUAPORE"	GUAPORE-CRENICICHLA	36
S23745	Crenicichla regani "TAPAJOS"	TAPAJOS-CRENICICHLA	37
S23748	Crenicichla regani sp.	BLACK-LINE-CRENICICHLA	37
S23750	Crenicichla reticulata	RETICULATA-CRENICICHLA	37
S23755	Crenicichla sp. "RIO-BRANCO"	RIO-BRANCO-CRENICICHLA	37
S23758	Crenicichla saxatilis	SAXA-CRENICICHLA	38
S23760	Crenicichla saxatilis "SURINAM"	SAXA-SURINAM-CRENICICHLA	37,38
S23765	Crenicichla saxatilis cf.	PEARL-CRENICICHLA	38
S23775	Crenicichla saxatilis sp.	BIG-POINT-CRENICICHLA	38
S23780	Crenicichla semifasciata	SEMI-CRENICICHLA	39,40
S23785	Crenicichla sp. "SINOP"	SINOP-CRENICICHLA	40
S23790	Crenicichla stocki	STOCKI-CRENICICHLA	40
S23795	Crenicichla strigata	STRIGATA-CRENICICHLA	40
S23800	Crenicichla sp. "SURINAM"	SURINAM-CRENICICHLA	41
S23805	Crenicichla sveni	META-CRENICICHLA	41
S23810	Crenicichla sveni cf.	GAITAN-CRENICICHLA	41
S23820	Crenicichla sp. "TAPAJOS" I	TAPAJOS I-CRENICICHLA	13,41

© A.C.S. Glaser GmbH

INDEX
Code - numbers

S23825	Crenicichla sp. "TAPAJOS"	TAPAJOS-CRENICICHLA	43
S23830	Crenicichla ternetzi	TERNETZI-CRENICICHLA	43
S23835	Crenicichla tigrina	TIGRINA-CRENICICHLA	43
S23845	Crenicichla urosema	UROSEMA CRENICICHLA	43
S23850	Crenicichla sp. "URUBAXI"	URUBAXI-CRENICICHLA	43
S23860	Crenicichla sp. "VENEZUELA" (befor strigata)	VENEZUELA-CRENICICHLA	43,44
S23863	Crenicichla sp. "VENEZUELA"	JUMBO-CRENICICHLA	44
S23865	Crenicichla sp. "VENEZUELA-DWARF"	VENEZUELA-DWARF-Crenicichla	44
S23870	Crenicichla vittata	VITTATA-CRENICICHLA	44,45
S23880	Crenicichla wallacii (?)	WALLACII-CRENICICHLA	45
S23885	Crenicichla wallacii cf.	AYACUCHO-CRENICICHLA	45
S23895	Crenicichla sp. "XINGU" I	XINGU I-CRENICICHLA	42,45
S23900	Crenicichla sp. "XINGU" II	XINGU II-CRENICICHLA	45,46
S23905	Crenicichla sp. "XINGU" III	XINGU III-CRENICICHLA	46
S23910	Crenicichla sp. "XINGU-ORANGE"	ORANGE-CRENICICHLA	46
S32105	Geophagus sp. "ALTAMIRA"	ALTAMIRA-GEOPHAGUS	48
S32110	Geophagus altifrons	ALTIFRONS-GEOPHAGUS	48
S32115	Geophagus altifrons "RIO-NEGRO"	RIO-NEGRO-ALTIFRONS	48
S32120	Geophagus altifrons "TAPAJOS"	TAPAJOS-ALTIFRONS	48
S32125	Geophagus altifrons "XINGU"	XINGU-ALTIFRONS	48
S32130	Geophagus altifrons cf.	ALTIFRONS-GEOPHAGUS	48
S32135	Geophagus sp. "ARAGUAIA"	ARAGUAIA-GEOPHAGUS	48
S32140	Geophagus sp. "AREÖES"	AREÖES-GEOPHAGUS	47,49
S32143	Geophagus sp. "AREÖES" 2-spot	TWO-SPOT-GEOPHAGUS	49
S32145	Geophagus argyrostictus	ARGY-GEOPHAGUS	49,50
S32150	Geophagus australis	ARGENTINA-GEOPHAGUAS	49
S32160	Geophagus brasiliensis	BRAZIL-GEOPHAGUS	51
S32165	Geophagus brasiliensis "FORM A"	BRAZIL A-GEOPHAGUS	51
S32170	Geophagus brasiliensis "FORM B"	BRAZIL B-GEOPHAGUS	51
S32175	Geophagus brasiliensis sp. cf.	BAHIA-GEOPHAGUS	51
S32185	Geophagus sp. "CAQUETA"	CAQUETA-GEOPHAGUS	52
S32190	Geophagus sp. "CARONI"	CARONI-GEOPHAGUS	52
S32195	Geophagus crassilabris	PANAMA-GEOPHAGUS	52
S32200	Geophagus sp. "CURUA"	CURUA-GEOPHAGUS	52
S32210	Geophagus grammepareius	CARONI-GEOPHAGUS	52
S32225	Geophagus hondae-steindachneri	HUMPHEAD-GEOPHAGUS	53
S32235	Geophagus sp. "INIRIDA"	INIRIDA-GEOPHAGUS	53
S32250	Geophagus sp. "MARABA" (cf.altifrons)	MARABA-GEOPHAGUS	53
S32260	Geophagus sp. "MITU"	MITU-GEOPHAGUS	53
S32275	Geophagus pellegrini	CHOCO-GEOPHAGUS	53
S32280	Geophagus sp. "PINDARE"	PINDARE-GEOPHAGUS	56
S32285	Geophagus proximus	AMAZONAS-GEOPHAGUS	54,56
S32295	Geophagus sp. "RED-PERU"	RED-PERU-GEOPHAGUS	56
S32305	Geophagus sp. "RIO-MOJU"	RIO-MOJU-GEOPHAGUS	56
S32310	Geophagus sp. "RIO-NEGRO" II	RIO-NEGRO II-Geophagus	57
S32320	Geophagus steindachneri (G. hondae)	COLUMBIA-GEOPHAGUS	57
S32330	Geophagus surinamensis	SURINAM-GEOPHAGUS	57
S32335	Geophagus surinamensis sp. "larvophil"	SURINAM-GEOPHAGUS	58
S32345	Geophagus surinamensis cf. "ovophil"	SURINAM-GEOPHAGUS	57,58
S32360	Geophagus taeniopareius	VENEZUELA-GEOPHAGUS	58
S32365	Geophagus sp. "TAPAJOS" I "RED-CHEEK"	RED-CHEEK-GEOPHAGUS	58
S32370	Geophagus sp. "TAPAJOS" II	TAPAJOS II-GEOPHAGUS	60
S32375	Geophagus sp. "TROMBETAS"	TROMBETAS-GEOPHAGUS	60
S32385	Geophagus sp. "VENEZUELA"	VENEZUELA-GEOPHAGUS	60
S34000	Guianacara sp. "CARONI"	CARONI-GUIANACARA	60
S34020	Guianacara owroewefi	BLUE-GUIANACARA	60
S34025	Guianacara owroewefi cf.	SHOCO-GUIANACARA	60
S34035	Guianacara sp. "RED-CHEEK"	RED-CHEEK-GUIANACARA	61
S34205	Gymnogeophagus sp. "BAHIA-RED"	BAHIA-GEOPHAGUS	61
S34210	Gymnogeophagus balzani	BALZANI-CICHLID	61,62,64
S34215	Gymnogeophagus sp. "BRAZIL" I	BRAZIL I-GEOPHAGUS	62
S34220	Gymnogeophagus sp. "BRAZIL" II	BRAZIL II-GEOPHAGUS	62

INDEX
Code - numbers

Code	Scientific name	Common name	Page
S34230	Gymnogeophagus gymnogenys	REDFIN-GEOPHAGUS	62,65
S34240	Gymnogeophagus labiatus	LABIATUS-GEOPHAGUS	62,63
S34245	Gymnogeophagus labiatus sp.	ROUNDHEAD-GEOPHAGUS	63
S34255	Gymnogeophagus meridionalis	MERIDIO-GEOPHAGUS	63
S34265	Gymnogeophagus rhabdotus	RHABDOTUS-GEOPHAGUS	63
S34275	Gymnogeophagus setequedas	QUEDAS-GEOPHAGUS	66
S78205	Retroculus lapidifer	MARBLE-RETROCULUS	66
S78215	Retroculus xinguensis	XINGU-RETROCULUS	66
S85305	Satanoperca acuticeps	FOUR-SPOT-GEOPHAGUS	67
S85315	Satanoperca sp. "COLUMBIA"	COLUMBIA-SATANOPERCA	67
S85325	Satanoperca daemon	THREE-SPOT-GEOPHAGUS	67,69
S85335	Satanoperca sp. "French-GUAYANA"	FRENCH-GUAYANA-SATANOPERCA	67
S85350	Satanoperca jurupari	EARTH-EATER	67,68
S85355	Satanoperca jurupari sp.	EARTH-EATER	68
S85365	Satanoperca leucosticta	META-SATANOPERCA	68
S85370	Satanoperca leucosticta cf.	PEARL-SATANOPERCA	71
S85375	Satanoperca lilith	ONE-SPOT-GEOPHAGUS	71
S85385	Satanoperca mapiritensis	ORINOCO-SATANOPERCA	71
S85395	Satanoperca pappaterra	PANTANAL-SATANOPERCA	71
S85405	Satanoperca sp. "REDLIP"	REDLIP-SATANOPERCA	72
S85410	Satanoperca sp. "RIO-XINGU" (lilith??)	XINGU-SATANOPERCA	72
S85415	Satanoperca sp. "Guayana"	GUAYANA-SATANOPERCA	72
S85420	Satanoperca sp. "UAUPES"	UAUPES-SATANOPERCA	72
S93004	Teleocichla sp. "PARU"	PARU-TELEO	72
S93005	Teleocichla I sp.	I-TELEO	73
S93010	Teleocichla II sp. "BLACK"	BLACK-TELEO II	73
S93015	Teleocichla III sp. "DOT"	DOT-TELEO III	73
S93020	Teleocichla IV sp. "GOBIO"	GOBIO-TELEO IV	73
S93030	Teleocichla centrarchus	CENTRO-TELEO	74
S93035	Teleocichla cinderella	CINDI-TELEO	74
S93040	Teleocichla cinderella cf.	BROADBAND-TELEO	74
S93050	Teleocichla gephyrogramma	GEPHYRO-TELEO	74
S93060	Teleocichla monogramma	MONOGRAMMA-TELEO	75
S93063	Teleocichla monogramma sp.	MONOGRAMMA-TELEO	75
S93070	Teleocichla prionogenys	PRIONO-TELEO	75
S93075	Teleocichla proselytus	LYTO-TELEO	75
S93085	Teleocichla sp.	SPOTLINE-TELEO	76
S93090	Teleocichla sp. "XINGU" II	XINGU II-TELEO	76
S93095	Teleocichla sp. "XINGU" III	XINGU III-TELEO	76
S93100	Teleocichla sp. "XINGU" IV	XINGU IV-TELEO	76
S99755	Uaru amphiacanthoides	WEDGESPOT-CICHLID	77

ACHTUNG / *ATTENTION*:

Einige Arten jetzt umbenannt:
Some species are now newly named:

Satanoperca	vorher / *before*	Geophagus
Gymnogeophagus	vorher / *before*	Geophagus
Guianacara geyai	vorher / *before*	Acarichthys geayi

© A.C.S. Glaser GmbH

INDEX
Alphabet

Acarichthy geayi (Guianacara geyai)	GEAYI-ACARICHTHYS	78
Acarichthys heckelii	HECKEL-ACARICHTHS	78
acuticeps Satanoperca	FOUR-SPOT-GEOPHAGUS	67
acutirostris Crenicichla	ACUTIS-CRENICICHLA	9
albopunctata Crenicichla	SPOTTED-CRENICICHLA	10
ALTA sp. Crenicichla	ALTA-CRENICICHLA	10
ALTAMIRA sp. Geophagus	ALTAMIRA-GEOPHAGUS	48
altifrons cf. Geophagus	ALTIFRONS-GEOPHAGUS	48
altifrons Geophagus	ALTIFRONS-GEOPHAGUS	48
amphiacanthoides Uaru	WEDGESPOT-CICHLID	77
anthurus Crenicichla	ANTHURUS-CRENICICHLA	10
APPROUAGE sp. Crenicichla	APPROUAGE-CRENICICHLA	10
ARAGUAIA sp. Geophagus	ARAGUAIA-GEOPHAGUS	48
ARAPIUNS sp. Crenicichla	ARAPIUNS-CRENICICHLA	11
AREÖES 2-spot sp. Geophagus	TWO-SPOT-GEOPHAGUS	49
AREÖES sp. Geophagus	AREÖES-GEOPHAGUS	47,49
argyrostictus Geophagus	ARGY-GEOPHAGUS	49,50
Astronotus ocellatus	OSCAR-WILDFORM	78
Astronotus ocellatus "ALBINO"	ALBINO-OSCAR	78
Astronotus ocellatus sp. "MARBLE"	MARBLE-OSCAR	78
Astronotus ocellatus sp. "PLANE-COLOUR"	BLACKFIN-OSCAR	79
Astronotus ocellatus sp. "RED-PEARL"	RED-PEARL-OSCAR	79
Astronotus ocellatus WILDFORM-BIGSPOT	OSKAR WILD-FORM	78
ATABAPO sp. Crenicichla	ATABAPO-CRENICICHLA	11
ATAPAPO lugubris cf. Crenicichla	ATABAPO-CRENICICHLA	23
australis Geophagus	ARGENTINA-GEOPHAGUAS	49
BAHIA-RED sp. Gymnogeophagus	BAHIA-GEOPHAGUS	61
balzani Gymnogeophagus	BALZANI-CICHLID	61,62,64
BELEM sp. Crenicichla	BELEM-PIKE-CICHLID	11
BELLY-CRAWLER sp. Crenicichla	BELLY-CRAWLER-CRENICICHLA	11
Biotodoma cupido	CUPIDO-CICHLID	79
Biotodoma cupido cf.	BELEM-BIOTODOMA	79
Biotodoma wavrini	WAVRINI-BIOTODOMA	79
Biotodoma wavrini cf.	ORINOCO-BIOTODOMA	79,80
bitaeniatus Chaetobranchopsis	TWO-SPOT-CHAETOBRANCHOPSIS	80
BLACK II sp. Teleocichla	BLACK-TELEO II	73
BOCON sp. Crenicichla	BOCON-CRENICICHLA	11
brasiliensis Geophagus	BRAZIL-GEOPHAGUS	51
brasiliensis sp. cf. Geophagus	BAHIA-GEOPHAGUS	51
BRAZIL I sp. Gymnogeophagus	BRAZIL I-GEOPHAGUS	62
BRAZIL II sp. Gymnogeophagus	BRAZIL II-GEOPHAGUS	62
cametana Crenicichla	TOCANTINS-CRENICICHLA	14
CAQUETA sp. Geophagus	CAQUETA-GEOPHAGUS	52
cardiostigma Crenicichla	RIO-BRANCO-CRENICICHLA	14
cardiostigma Crenicichla	PTEROGRAMMA-CRENICICHLA	14
CARONI sp. Geophagus	CARONI-GEOPHAGUS	52
CARONI sp. Guianacara	CARONI-GUIANACARA	60
CASIQUIARE sp. Crenicichla	CASIQUIARE-CRENICICHLA	14
centrarchus Teleocichla	CENTRO-TELEO	74
Chaetobranchopsis bitaeniatus	TWO-SPOT-CHAETOBRANCHOPSIS	80
Chaetobranchopsis flavescens	FLAVI-CHAETOBRANCHOPSIS	80
Chaetobranchopsis orbicularis	ORBI-CHAETOBRANCHOPSIS	80
Chaetobranchopsis orbicularis sp.	BLACK-BAND-CHAETOBRANCHOPSIS	80
Chaetobranchopsis spectabilis	SPECTA-CHAETOBRANCHOPSIS	80
Cichla ocellaris	PIKE-CICHLID	7,12
Cichla ocellaris cf. "RIO PRETO"	PRETO-PIKE-CICHLID	7
Cichla orinocensis	ORINOCO-PIKE-CICHLID	7
Cichla orinocensis cf. "RIO-INIRIDA"	INIRIDA-PIKE-CICHLID	8
Cichla orinocensis cf. "RIO-VAUPES"	VAUPES-PIKE-CICHLID	8
Cichla sp. "ARAGUAIA" III	ARAGUAIA-CICHLA	7
Cichla sp. "TAPAJOS" I	TAPAJOS-CICHLA	8
Cichla sp. "XINGU" II	XINGU II-CICHLA	9

INDEX
Alphabet

Cichla temensis	TEMENSIS-PIKE-CICHLID	8
Cichla temensis cf. "ARAGUAIA"	ARAGUAIA-CICHLA	9
Cichla temensis cf. "CARONI"	CARONI-CICHLA	8
Cichla temensis cf. "CAURA"	CAURA-CICHLA	9
Cichla temensis cf. "VAUPES"	VAUPES-CICHLA	9
cincta Crenicichla	CINCTA-CRENICICHLA	14
cinderella cf. Teleocichla	BROADBAND-TELEO	74
cinderella Teleocichla	CINDI-TELEO	74
COLUMBIA sp. Satanoperca	COLUMBIA-SATANOPERCA	67
compressiceps Crenicichla	COMPRESS-CRENICICHLA	15,21
crassilabris Geophagus	PANAMA-GEOPHAGUS	52
Crenicichla "ORINOCO"	ORINOCO-CRENICICHLA	31
Crenicichla acutirostris	ACUTIS-CRENICICHLA	9
Crenicichla albopunctata	SPOTTED-CRENICICHLA	10
Crenicichla anthurus	ANTHURUS-CRENICICHLA	10
Crenicichla cametana	TOCANTINS-CRENICICHLA	14
Crenicichla cardiostigma	RIO-BRANCO-CRENICICHLA	14
Crenicichla cardiostigma (pterogramma?)	PTEROGRAMMA-CRENICICHLA	14
Crenicichla cincta	CINCTA-CRENICICHLA	14
Crenicichla compressiceps	COMPRESS-CRENICICHLA	15,21
Crenicichla cyanonotus	CYANO-CRENICICHLA	15
Crenicichla cyclostoma	CYCLOS-CRENICICHLA	15
Crenicichla dorsocellata	CAMPOS-CRENICICHLA	15
Crenicichla edithae	EDITH-CRENICICHLA	16
Crenicichla frenata	TRINIDAD-CRENICICHLA	16
Crenicichla frenata var.	FRENATA-CRENICICHLA	16
Crenicichla geayi	GEAYI-CRENICICHLA	16
Crenicichla geayi cf.	BRIGHT-BLUE-CRENICICHLA	17
Crenicichla heckeli	HECKEL-CRENICICHLA	18
Crenicichla jaguarensis	JAGUAR-CRENICICHLA	19
Crenicichla jegui	JEGUI-CRENICICHLA	19
Crenicichla jegui sp.	NORTH-EAST-CRENICICHLA	20
Crenicichla johanna	JOHANNA-CRENICICHLA	20
Crenicichla johanna var. cf.	JOHANNA-VAR.-CRENICICHLA	20
Crenicichla labrina	LABRINA-CRENICICHLA	20
Crenicichla labrina sp.	AMAZONAS-CRENICICHLA	20
Crenicichla lenticulata	LENTICULATA-CRENICICHLA	22
Crenicichla lepidota	LEPIDOTA-CRENICICHLA	22
Crenicichla lepidota "REAL"	REAL-LEPIDOTA-CRENICICHLA	22
Crenicichla lucius	LUCIUS-CRENICICHLA	22,23
Crenicichla lugubris	LUGU-CRENICICHLA	23,24
Crenicichla lugubris cf. "ATABAPO"	ATABAPO-CRENICICHLA	23
Crenicichla lugubris var.	RED-MOUTH-CRENICICHLA	23
Crenicichla macrophthalma	MACRO-CRENICICHLA	26
Crenicichla marmorata	MARBLE-CRENICICHLA	26,27
Crenicichla menezesi sp.	MENEZESI-CRENICICHLA	27,28
Crenicichla multispinosa	MULTI-CRENICICHLA	28
Crenicichla nothophthalmus sp.	NEGRO-CRENICICHLA	28
Crenicichla nothophthalmus sp. "UAUPES"	UAUPES-CRENICICHLA	31
Crenicichla notophthalmus	NORTH-CRENICICHLA	25,28
Crenicichla notophthalmus cf.	BIG-POINT-CRENICICHLA	28
Crenicichla percna	PERCNA-CRENICICHLA	32
Crenicichla phaiospilus	SPILUS-CRENICICHLA	30,32
Crenicichla proteus	PERU-CRENICICHLA	33
Crenicichla punctata	PUNCTATA-CRENICICHLA	33
Crenicichla regani	REGANI-CRENICICHLA	35
Crenicichla regani "DAS MORTES"	MORTES-CRENICICHLA	34,35
Crenicichla regani "GUAPORE"	GUAPORE-CRENICICHLA	36
Crenicichla regani "RIO-ACARA"	ACARA-CRENICICHLA	35
Crenicichla regani "RIO-GUAMA"	GUAMA-CRENICICHLA	35
Crenicichla regani "RIO-NEGRO"	RIO-NEGRO-CRENICICHLA	34,36
Crenicichla regani "TAPAJOS"	TAPAJOS-CRENICICHLA	37

INDEX
Alphabet

Crenicichla regani "TEFE"	TEFE-CRENICICHLA	36
Crenicichla regani "TROMBETAS"	TROMBETAS-CRENICICHLA	36
Crenicichla regani sp.	BLACK-LINE-CRENICICHLA	37
Crenicichla reticulata	RETICULATA-CRENICICHLA	37
Crenicichla saxatilis	SAXA-CRENICICHLA	38
Crenicichla saxatilis "SURINAM"	SAXA-SURINAM-CRENICICHLA	37,38
Crenicichla saxatilis cf.	PEARL-CRENICICHLA	38
Crenicichla saxatilis sp.	BIG-POINT-CRENICICHLA	38
Crenicichla semifasciata	SEMI-CRENICICHLA	39,40
Crenicichla sp. "ALTA"	ALTA-CRENICICHLA	10
Crenicichla sp. "APPROUAGE"	APPROUAGE-CRENICICHLA	10
Crenicichla sp. "ARAPIUNS"	ARAPIUNS-CRENICICHLA	11
Crenicichla sp. "ATABAPO"	ATABAPO-CRENICICHLA	11
Crenicichla sp. "BELEM"	BELEM-PIKE-CICHLID	11
Crenicichla sp. "BELLY-CRAWLER"	BELLY-CRAWLER-CRENICICHLA	11
Crenicichla sp. "BOCON"	BOCON-CRENICICHLA	11
Crenicichla sp. "CASIQUIARE"	CASIQUIARE-CRENICICHLA	14
Crenicichla sp. "GUANIAMO"	GUANIAMO-CRENICICHLA	17
Crenicichla sp. "GUARIQUITO"	GUARIQUITO-CRENICICHLA	17
Crenicichla sp. "GUARIQUITO" var.	AYACUCHO-CRENICICHLA	17,18
Crenicichla sp. "GUAYANA"	DEMERARA-CRENICICHLA	18
Crenicichla sp. "INIRIDA" I	INIRIDA I-CRENICICHLA	18
Crenicichla sp. "INIRIDA" II	INIRIDA II-CRENICICHLA	18
Crenicichla sp. "INIRIDA" III "RED-SPOT"	INIRIDA III-"RED-SPOT"-Crenicichla	19
Crenicichla sp. "JABUTI"	JABUTI-CRENICICHLA	19
Crenicichla sp. "MAICURU"	MAICURU-CRENICICHLA	26
Crenicichla sp. "MANAUS"	MANAUS-CRENICICHLA	26
Crenicichla sp. "MARMORATA"	MARMORATA-CRENICICHLA	27
Crenicichla sp. "MATO-GROSSO"	MATO-GROSSO-CRENICICHLA	27
Crenicichla sp. "ORANGE"-Xingu/Belem	ORANGE-XINGU-CRENICICHLA	31
Crenicichla sp. "ORINOCO"	MANY-SPOT-CRENICICHLA	31
Crenicichla sp. "PACAYA"	PACAYA-CRENICICHLA	31
Crenicichla sp. "PERNAMBUCO"	PERNAMBUCO-CRENICICHLA	32
Crenicichla sp. "PINDARE"	PINDARE-CRENICICHLA	33
Crenicichla sp. "RED-BELLY"	RED-BELLY-Crenicichla	33
Crenicichla sp. "RIO-BRANCO"	RIO-BRANCO-CRENICICHLA	37
Crenicichla sp. "SINOP"	SINOP-CRENICICHLA	40
Crenicichla sp. "SURINAM"	SURINAM-CRENICICHLA	41
Crenicichla sp. "TAPAJOS"	TAPAJOS-CRENICICHLA	43
Crenicichla sp. "TAPAJOS" I	TAPAJOS I-CRENICICHLA	13,41
Crenicichla sp. "URUBAXI"	URUBAXI-CRENICICHLA	43
Crenicichla sp. "VENEZUELA"	JUMBO-CRENICICHLA	44
Crenicichla sp. "VENEZUELA" (befor strigata)	VENEZUELA-CRENICICHLA	43,44
Crenicichla sp. "VENEZUELA-DWARF"	VENEZUELA-DWARF-Crenicichla	44
Crenicichla sp. "XINGU" I	XINGU I-CRENICICHLA	42,45
Crenicichla sp. "XINGU" II	XINGU II-CRENICICHLA	45,46
Crenicichla sp. "XINGU" III	XINGU III-CRENICICHLA	46
Crenicichla sp. "XINGU-ORANGE"	ORANGE-CRENICICHLA	46
Crenicichla stocki	STOCKI-CRENICICHLA	40
Crenicichla strigata	STRIGATA-CRENICICHLA	40
Crenicichla sveni	META-CRENICICHLA	41
Crenicichla sveni cf.	GAITAN-CRENICICHLA	41
Crenicichla ternetzi	TERNETZI-CRENICICHLA	43
Crenicichla tigrina	TIGRINA-CRENICICHLA	43
Crenicichla urosema	UROSEMA CRENICICHLA	43
Crenicichla vittata	VITTATA-CRENICICHLA	44,45
Crenicichla wallacii (?)	WALLACII-CRENICICHLA	45
Crenicichla wallacii cf.	AYACUCHO-CRENICICHLA	45
cupido Biotodoma	CUPIDO-CICHLID	79
cupido cf. Biotodoma	BELEM-BIOTODOMA	79
CURUA sp. Geophagus	CURUA-GEOPHAGUS	52
cyanonotus Crenicichla	CYANO-CRENICICHLA	15

INDEX
Alphabet

cyclostoma Crenicichla	CYCLOS-CRENICICHLA	15
daemon Satanoperca	THREE-SPOT-GEOPHAGUS	67,69
DAS MORTES regani Crenicichla	MORTES-CRENICICHLA	34,35
dorsocellata Crenicichla	CAMPOS-CRENICICHLA	15
DOT III sp. Teleocichla	DOT-TELEO III	73
edithae Crenicichla	EDITH-CRENICICHLA	16
flavescens Chaetobranchopsis	FLAVI-CHAETOBRANCHOPSIS	80
FORM A brasiliensis Geophagus	BRAZIL A-GEOPHAGUS	51
FORM B brasiliensis Geophagus	BRAZIL B-GEOPHAGUS	51
frenata Crenicichla	TRINIDAD-CRENICICHLA	16
frenata var. Crenicichla	FRENATA-CRENICICHLA	16
FRENCH-GUAYANA sp. Satanoperca	FRENCH-GUAYANA-SATANOPERCA	67
geayi Acarichthys (geayi Guianacara)	GEAYI-ACARICHTHYS	78
geayi cf. Crenicichla	BRIGHT-BLUE-CRENICICHLA	17
geayi Crenicichla	GEAYI-CRENICICHLA	16
Geophagus altifrons	ALTIFRONS-GEOPHAGUS	48
Geophagus altifrons "RIO-NEGRO"	RIO-NEGRO-ALTIFRONS	48
Geophagus altifrons "TAPAJOS"	TAPAJOS-ALTIFRONS	48
Geophagus altifrons "XINGU"	XINGU-ALTIFRONS	48
Geophagus altifrons cf.	ALTIFRONS-GEOPHAGUS	48
Geophagus argyrostictus	ARGY-GEOPHAGUS	49,50
Geophagus australis	ARGENTINA-GEOPHAGUAS	49
Geophagus brasiliensis	BRAZIL-GEOPHAGUS	51
Geophagus brasiliensis "FORM A"	BRAZIL A-GEOPHAGUS	51
Geophagus brasiliensis "FORM B"	BRAZIL B-GEOPHAGUS	51
Geophagus brasiliensis sp. cf.	BAHIA-GEOPHAGUS	51
Geophagus crassilabris	PANAMA-GEOPHAGUS	52
Geophagus grammepareius	CARONI-GEOPHAGUS	52
Geophagus hondae-steindachneri	HUMPHEAD-GEOPHAGUS	53
Geophagus pellegrini	CHOCO-GEOPHAGUS	53
Geophagus proximus	AMAZONAS-GEOPHAGUS	54,56
Geophagus sp. "ALTAMIRA"	ALTAMIRA-GEOPHAGUS	48
Geophagus sp. "ARAGUAIA"	ARAGUAIA-GEOPHAGUS	48
Geophagus sp. "AREÕES"	AREÕES-GEOPHAGUS	47,49
Geophagus sp. "AREÕES" 2-spot	TWO-SPOT-GEOPHAGUS	49
Geophagus sp. "CAQUETA"	CAQUETA-GEOPHAGUS	52
Geophagus sp. "CARONI"	CARONI-GEOPHAGUS	52
Geophagus sp. "CURUA"	CURUA-GEOPHAGUS	52
Geophagus sp. "INIRIDA"	INIRIDA-GEOPHAGUS	53
Geophagus sp. "MARABA" (cf.altifrons)	MARABA-GEOPHAGUS	53
Geophagus sp. "MITU"	MITU-GEOPHAGUS	53
Geophagus sp. "PINDARE"	PINDARE-GEOPHAGUS	56
Geophagus sp. "RED-PERU"	RED-PERU-GEOPHAGUS	56
Geophagus sp. "RIO-MOJU"	RIO-MOJU-GEOPHAGUS	56
Geophagus sp. "RIO-NEGRO" II	RIO-NEGRO II-Geophagus	57
Geophagus sp. "TAPAJOS" I"RED-CHEEK"	RED-CHEEK-GEOPHAGUS	58
Geophagus sp. "TAPAJOS" II	TAPAJOS II-GEOPHAGUS	60
Geophagus sp. "TROMBETAS"	TROMBETAS-GEOPHAGUS	60
Geophagus sp. "VENEZUELA"	VENEZUELA-GEOPHAGUS	60
Geophagus steindachneri (G. hondae)	COLUMBIA-GEOPHAGUS	57
Geophagus surinamensis	SURINAM-GEOPHAGUS	57
Geophagus surinamensis cf. "ovophil"	SURINAM-GEOPHAGUS	57,58
Geophagus surinamensis sp. "larvophil"	SURINAM-GEOPHAGUS	58
Geophagus taeniopareius	VENEZUELA-GEOPHAGUS	58
gephyrogramma Teleocichla	GEPHYRO-TELEO	74
GOBIO IV sp. Teleocichla	GOBIO-TELEO IV	73
grammepareius Geophagus	CARONI-GEOPHAGUS	52
GUANIAMO sp. Crenicichla	GUANIAMO-CRENICICHLA	17
GUAPORE regani Crenicichla	GUAPORE-CRENICICHLA	36
GUARIQUITO sp. Crenicichla	GUARIQUITO-CRENICICHLA	17
GUARIQUITO sp. var. Crenicichla	AYACUCHO-CRENICICHLA	17,18
GUAYANA sp. Crenicichla	DEMERARA-CRENICICHLA	18

INDEX
Alphabet

GUAYANA sp. Satanoperca	GUAYANA-SATANOPERCA	72
Guianacara owroewefi	BLUE-GUIANACARA	60
Guianacara owroewefi cf.	SHOCO-GUIANACARA	60
Guianacara sp. "CARONI"	CARONI-GUIANACARA	60
Guianacara sp. "RED-CHEEK"	RED-CHEEK-GUIANACARA	61
gymnogenys Gymnogeophagus	REDFIN-GEOPHAGUS	62,65
Gymnogeophagus balzani	BALZANI-CICHLID	61,62,64
Gymnogeophagus gymnogenys	REDFIN-GEOPHAGUS	62,65
Gymnogeophagus labiatus	LABIATUS-GEOPHAGUS	62,63
Gymnogeophagus labiatus sp.	ROUNDHEAD-GEOPHAGUS	63
Gymnogeophagus meridionalis	MERIDIO-GEOPHAGUS	63
Gymnogeophagus rhabdotus	RHABDOTUS-GEOPHAGUS	63
Gymnogeophagus setequedas	QUEDAS-GEOPHAGUS	66
Gymnogeophagus sp. "BAHIA-RED"	BAHIA-GEOPHAGUS	61
Gymnogeophagus sp. "BRAZIL" I	BRAZIL I-GEOPHAGUS	62
Gymnogeophagus sp. "BRAZIL" II	BRAZIL II-GEOPHAGUS	62
heckeli Crenicichla	HECKEL-CRENICICHLA	18
heckelii Acarichthys	HECKEL-ACARICHTHS	78
hondae-steindachneri Geophagus	HUMPHEAD-GEOPHAGUS	53
I sp. Teleocichla	I-TELEO	73
INIRIDA I sp. Crenicichla	INIRIDA I-CRENICICHLA	18
INIRIDA II sp. Crenicichla	INIRIDA II-CRENICICHLA	18
INIRIDA III sp. Crenicichla RED-SPOT	INIRIDA III-"RED-SPOT"-Crenicichla	19
INIRIDA sp. Geophagus	INIRIDA-GEOPHAGUS	53
JABUTI sp. Crenicichla	JABUTI-CRENICICHLA	19
jaguarensis Crenicichla	JAGUAR-CRENICICHLA	19
jegui Crenicichla	JEGUI-CRENICICHLA	19
jegui sp. Crenicichla	NORTH-EAST-CRENICICHLA	20
johanna cf. var. Crenicichla	JOHANNA-VAR.-CRENICICHLA	20
johanna Crenicichla	JOHANNA-CRENICICHLA	20
jurupari Satanoperca	EARTH-EATER	67,68
jurupari sp. Satanoperca	EARTH-EATER	68
labiatus Gymnogeophagus	LABIATUS-GEOPHAGUS	62,63
labiatus sp. Gymnogeophagus	ROUNDHEAD-GEOPHAGUS	63
labrina Crenicichla	LABRINA-CRENICICHLA	20
labrina sp. Crenicichla	AMAZONAS-CRENICICHLA	20
lapidifer Retroculus	MARBLE-RETROCULUS	66
lenticulata Crenicichla	LENTICULATA-CRENICICHLA	22
lepidota Crenicichla	LEPIDOTA-CRENICICHLA	22
leucosticta cf. Satanoperca	PEARL-SATANOPERCA	71
leucosticta Satanoperca	META-SATANOPERCA	68
lilith Satanoperca	ONE-SPOT-GEOPHAGUS	71
lucius Crenicichla	LUCIUS-CRENICICHLA	22,23
lugubris Crenicichla	LUGU-CRENICICHLA	23,24
lugubris var. Crenicichla	RED-MOUTH-CRENICICHLA	23
macrophthalma Crenicichla	MACRO-CRENICICHLA	26
MAICURU sp. Crenicichla	MAICURU-CRENICICHLA	26
MANAUS sp. Crenicichla	MANAUS-CRENICICHLA	26
mapiritensis Satanoperca	ORINOCO-SATANOPERCA	71
MARABA (cf. altifrons) sp. Geophagus	MARABA-GEOPHAGUS	53
marmorata Crenicichla	MARBLE-CRENICICHLA	26,27
MARMORATA sp. Crenicichla	MARMORATA-CRENICICHLA	27
MATO-GROSSO sp. Crenicichla	MATO-GROSSO-CRENICICHLA	27
menezesi sp. Crenicichla	MENEZESI-CRENICICHLA	27,28
meridionalis Gymnogeophagus	MERIDIO-GEOPHAGUS	63
MITU sp. Geophagus	MITU-GEOPHAGUS	53
monogramma sp. Teleocichla	MONOGRAMMA-TELEO	75
monogramma Teleocichla	MONOGRAMMA-TELEO	75
multispinosa Crenicichla	MULTI-CRENICICHLA	28
notophthalmus cf. Crenicichla	BIG-POINT-CRENICICHLA	28
notophthalmus Crenicichla	NORTH-CRENICICHLA	25,28
notophthalmus sp. Crenicichla	NEGRO-CRENICICHLA	28

INDEX
Alphabet

ocellaris cf. *Cichla*	PRETO-PIKE-CICHLID	7
ocellaris, Cichla	PIKE-CICHLID	7,12
ocellatus ALBINO *Astronotus*	ALBINO-OSCAR	78
ocellatus Astronotus	OSCAR-WILDFORM	78
ocellatus sp. MARBLE *Astronotus*	MARBLE-OSCAR	78
ocellatus sp. PLANE-COLOUR *Astronotus*	BLACKFIN-OSCAR	79
ocellatus sp. RED-PEARL *Astronotus*	RED-PEARL-OSCAR	79
ocellatus WILDFORM-BIGSPOT *Astronotus*	OSKAR WILD-FORM	78
ORANGE-Xingu/Belem sp. *Crenicichla*	ORANGE-XINGU-CRENICICHLA	31
orbicularis Chaetobranchopsis	ORBI-CHAETOBRANCHOPSIS	80
orbicularis sp. *Chaetobranchopsis*	BLACK-BAND-CHAETOBRANCHOPSIS	80
orinocensis cf. *Cichla*	VAUPES-PIKE-CICHLID	8
orinocensis cf. RIO-INIRIDA *Cichla*	INIRIDA-PIKE-CICHLID	8
orinocensis Cichla	ORINOCO-PIKE-CICHLID	7
ORINOCO *Crenicichla*	ORINOCO-CRENICICHLA	31
ORINOCO sp. *Crenicichla*	MANY-SPOT-CRENICICHLA	31
owroewefi cf. *Guianacara*	SHOCO-GUIANACARA	60
owroewefi Guianacara	BLUE-GUIANACARA	60
PACAYA sp. *Crenicichla*	PACAYA-CRENICICHLA	31
pappaterra Satanoperca	PANTANAL-SATANOPERCA	71
PARU sp. *Teleocichla*	PARU-TELEO	72
pellegrini Geophagus	CHOCO-GEOPHAGUS	53
percna Crenicichla	PERCNA-CRENICICHLA	32
PERNAMBUCO sp. *Crenicichla*	PERNAMBUCO-CRENICICHLA	32
phaiospilus Crenicichla	SPILUS-CRENICICHLA	30,32
PINDARE sp. *Crenicichla*	PINDARE-CRENICICHLA	33
PINDARE sp. *Geophagus*	PINDARE-GEOPHAGUS	56
prionogenys Teleocichla	PRIONO-TELEO	75
proselytus Teleocichla	LYTO-TELEO	75
proteus Crenicichla	PERU-CRENICICHLA	33
proximus Geophagus	AMAZONAS-GEOPHAGUS	54,56
punctata Crenicichla	PUNCTATA-CRENICICHLA	33
REAL *lepidota Crenicichla*	REAL-LEPIDOTA-CRENICICHLA	22
RED-BELLY sp. *Crenicichla*	RED-BELLY-Crenicichla	33
RED-CHEEK sp. *Guianacara*	RED-CHEEK-GUIANACARA	61
RED-PERU sp. *Geophagus*	RED-PERU-GEOPHAGUS	56
REDLIP sp. *Satanoperca*	REDLIP-SATANOPERCA	72
regani Crenicichla	REGANI-CRENICICHLA	35
regani sp. *Crenicichla*	BLACK-LINE-CRENICICHLA	37
reticulata Crenicichla	RETICULATA-CRENICICHLA	37
Retroculus lapidifer	MARBLE-RETROCULUS	66
Retroculus xinguensis	XINGU-RETROCULUS	66
rhabdotus Gymnogeophagus	RHABDOTUS-GEOPHAGUS	63
RIO-ACARA *regani Crenicichla*	ACARA-CRENICICHLA	35
RIO-BRANCO sp. *Crenicichla*	RIO-BRANCO-CRENICICHLA	37
RIO-GUAMA *regani Crenicichla*	GUAMA-CRENICICHLA	35
RIO-MOJU sp. *Geophagus*	RIO-MOJU-GEOPHAGUS	56
RIO-NEGRO *altifrons Geophagus*	RIO-NEGRO-ALTIFRONS	48
RIO-NEGRO II sp. *Geophagus*	RIO-NEGRO II-Geophagus	57
RIO-NEGRO *regani Crenicichla*	RIO-NEGRO-CRENICICHLA	34,36
RIO-XINGU sp. *Satanoperca (lilith??)*	XINGU-SATANOPERCA	72
Satanoperca acuticeps	FOUR-SPOT-GEOPHAGUS	67
Satanoperca daemon	THREE-SPOT-GEOPHAGUS	67,69
Satanoperca jurupari	EARTH-EATER	67,68
Satanoperca jurupari sp.	EARTH-EATER	68
Satanoperca leucosticta	META-SATANOPERCA	68
Satanoperca leucosticta cf.	PEARL-SATANOPERCA	71
Satanoperca lilith	ONE-SPOT-GEOPHAGUS	71
Satanoperca mapiritensis	ORINOCO-SATANOPERCA	71
Satanoperca pappaterra	PANTANAL-SATANOPERCA	71
Satanoperca sp. "COLUMBIA"	COLUMBIA-SATANOPERCA	67
Satanoperca sp. "French-GUAYANA"	FRENCH-GUAYANA-SATANOPERCA	67

INDEX
Alphabet

Satanoperca sp. "Guayana"	GUAYANA-SATANOPERCA	72
Satanoperca sp. "REDLIP"	REDLIP-SATANOPERCA	72
Satanoperca sp. "RIO-XINGU" (lilith??)	XINGU-SATANOPERCA	72
Satanoperca sp. "UAUPES"	UAUPES-SATANOPERCA	72
saxatilis cf. Crenicichla	PEARL-CRENICICHLA	38
saxatilis Crenicichla	SAXA-CRENICICHLA	38
saxatilis sp. Crenicichla	BIG-POINT-CRENICICHLA	38
semifasciata Crenicichla	SEMI-CRENICICHLA	39,40
setequedas Gymnogeophagus	QUEDAS-GEOPHAGUS	66
SINOP sp. Crenicichla	SINOP-CRENICICHLA	40
sp. I Cichla	TAPAJOS-CICHLA	8
sp. III ARAGUAIA Cichla	ARAGUAIA-CICHLA	7
sp. Teleocichla	SPOTLINE-TELEO	76
sp. XINGU II Cichla	XINGU II-CICHLA	9
spectabilis Chaetobranchopsis	SPECTA-CHAETOBRANCHOPSIS	80
steindachneri Geophagus (G. hondae)	COLUMBIA-GEOPHAGUS	57
stocki Crenicichla	STOCKI-CRENICICHLA	40
strigata Crenicichla	STRIGATA-CRENICICHLA	40
SURINAM saxatilis Crenicichla	SAXA-SURINAM-CRENICICHLA	37,38
SURINAM sp. Crenicichla	SURINAM-CRENICICHLA	41
surinamensis cf. "ovophil" Geophagus	SURINAM-GEOPHAGUS	57,58
surinamensis Geophagus	SURINAM-GEOPHAGUS	57
surinamensis sp. "larvophil" Geophagus	SURINAM-GEOPHAGUS	58
sveni cf. Crenicichla	GAITAN-CRENICICHLA	41
sveni Crenicichla	META-CRENICICHLA	41
taeniopareius Geophagus	VENEZUELA-GEOPHAGUS	58
TAPAJOS altifrons Geophagus	TAPAJOS-ALTIFRONS	48
TAPAJOS I "RED-CHEEK" sp. Geophagus	RED-CHEEK-GEOPHAGUS	58
TAPAJOS I sp. Crenicichla	TAPAJOS I-CRENICICHLA	13,41
TAPAJOS II sp. Geophagus	TAPAJOS II-GEOPHAGUS	60
TAPAJOS regani Crenicichla	TAPAJOS-CRENICICHLA	37
TAPAJOS sp. Crenicichla	TAPAJOS-CRENICICHLA	43
TEFE regani Crenicichla	TEFE-CRENICICHLA	36
Teleocichla centrarchus	CENTRO-TELEO	74
Teleocichla cinderella	CINDI-TELEO	74
Teleocichla cinderella cf.	BROADBAND-TELEO	74
Teleocichla gephyrogramma	GEPHYRO-TELEO	74
Teleocichla I sp.	I-TELEO	73
Teleocichla II sp. "BLACK"	BLACK-TELEO II	73
Teleocichla III sp. "DOT"	DOT-TELEO III	73
Teleocichla IV sp. "GOBIO"	GOBIO-TELEO IV	73
Teleocichla monogramma	MONOGRAMMA-TELEO	75
Teleocichla monogramma sp.	MONOGRAMMA-TELEO	75
Teleocichla prionogenys	PRIONO-TELEO	75
Teleocichla proselytus	LYTO-TELEO	75
Teleocichla sp.	SPOTLINE-TELEO	76
Teleocichla sp. "PARU"	PARU-TELEO	72
Teleocichla sp. "XINGU" II	XINGU II-TELEO	76
Teleocichla sp. "XINGU" III	XINGU III-TELEO	76
Teleocichla sp. "XINGU" IV	XINGU IV-TELEO	76
temensis cf. ARAGUAIA Cichla	ARAGUAIA-CICHLA	9
temensis cf. CARONI Cichla	CARONI-CICHLA	8
temensis cf. CAURA Cichla	CAURA-CICHLA	9
temensis cf. VAUPES Cichla	VAUPES-CICHLA	9
temensis Cichla	TEMENSIS-PIKE-CICHLID	8
ternetzi Crenicichla	TERNETZI-CRENICICHLA	43
tigrina Crenicichla	TIGRINA-CRENICICHLA	43
TROMBETAS regani Crenicichla	TROMBETAS-CRENICICHLA	36
TROMBETAS sp. Geophagus	TROMBETAS-GEOPHAGUS	60
Uaru amphiacanthoides	WEDGESPOT-CICHLID	77
UAUPES nothophthalmus sp. Crenicichla	UAUPES-CRENICICHLA	31
UAUPES sp. Satanoperca	UAUPES-SATANOPERCA	72

INDEX
Alphabet

urosema Crenicichla	UROSEMA CRENICICHLA	43
URUBAXI sp. Crenicichla	URUBAXI-CRENICICHLA	43
VENEZUELA sp. Crenicichla	JUMBO-CRENICICHLA	44
VENEZUELA sp. Crenicichla (befor strigata)	VENEZUELA-CRENICICHLA	43,44
VENEZUELA sp. Geophagus	VENEZUELA-GEOPHAGUS	60
VENEZUELA-DWARF sp. Crenicichla	VENEZUELA-DWARF-Crenicichla	44
vittata Crenicichla	VITTATA-CRENICICHLA	44,45
wallacii cf. Crenicichla	AYACUCHO-CRENICICHLA	45
wallacii Crenicichla	WALLACII-CRENICICHLA	45
wavrini Biotodoma	WAVRINI-BIOTODOMA	79
wavrini cf. Biotodoma	ORINOCO-BIOTODOMA	79,80
XINGU altifrons Geophagus	XINGU-ALTIFRONS	48
XINGU I sp. Crenicichla	XINGU I-CRENICICHLA	42,45
XINGU II sp. Crenicichla	XINGU II-CRENICICHLA	45,46
XINGU II sp. Teleocichla	XINGU II-TELEO	76
XINGU III sp. Crenicichla	XINGU III-CRENICICHLA	46
XINGU III sp. Teleocichla	XINGU III-TELEO	76
XINGU IV sp. Teleocichla	XINGU IV-TELEO	76
XINGU-ORANGE sp. Crenicichla	ORANGE-CRENICICHLA	46
xinguensis Retroculus	XINGU-RETROCULUS	66

ACHTUNG / *ATTENTION*:

Einige Arten jetzt umbenannt:
Some species are now newly named:

Satanoperca	vorher / *before*	Geophagus
Gymnogeophagus	vorher / *before*	Geophagus
Guianacara geyai	vorher / *before*	Acarichthys geayi

South America **Cichlids I**

Index der Bildautoren
index of the photographers

Bork

1x S32295, 1x S85335, 2x S85415

Mayland

1x S00106, 1x S06258, 1x S06261, 1x S06265, 1x S06268, 1x S07703, 1x S07715, 1x S11908, 1x S11908
2x S11915, 2x S13703, 1x S13718, 1x S13758, 1x Seite 21, 2x S23663, 3x S23735, 1x S23795, 1x S23895,
1x S23910, 1x S32110, 1x S32135, 1x Seite 47, 2x Seite 50, 2x S32165, 2x S32170, 1x S32275, 1x S32285,
1x Seite 54, 1x S32320, 1x S32330, 2x Seite 59, 2x S34025, 1x S34035, 1x S34210, 1x Seite 64, 1x S34230,
2x S34245, 1x S34265, 3x S78205, 1x S85305, 1x Seite 69, 2x S85355, 2x S85365, 1x S85395, 1x S85410,
1x S93040, 2x S93085, 2x S99755, 2x Seite 12

Mayland/Glaser

1x S23800, 2x S23865

Pinhard/Glaser

1x S23400, 1x S23475, 1x S23620, 1x S23625, 1x S23640, 1x S23670, 1x S23765, 1x S23865, 1x S23910,
1x S32285, 1x S32295, 1x S32305, 2x S32330, 1x S34210, 2x S34255, 3x S85350

Reinhard/Migge

1x S32150, 1x S32225, 1x S85365

Römer

1x S13725, 3x S23664, 1x S23880, 1x S23885, 1x S85420

Warzel/Minde

3x S23310, 2x S23315, 2x S23320, 2x S23325, 1x S23335, 1x S23340, 2x S23355, 2x S23360, 2x S23370,
2x S23378, 2x S23390, 2x S23395, 2x S23405, 1x S23415, 2x S23425, 2x S23435, 2x S23440, 2x S23450
2x S23470, 1x S23475, 2x S23480, 2x S23490, 1x S23525, 3x S23530, 1x S23535, 3x S23540, 1x S23545,
3x S23560, 3x S23570, 1x S23580, 2x S23590, 3x S23595, 2x S23596, 1x S23610, 1x S23615, 2x S23620,
4x S23625, 1x S23630, 2x S23640, 1x S23645, 2x S23655, 2x S23660, 2x S23680, 1x S23690, 2x S23695,
1x S23710, 2x S23720, 2x S23725, 2x S23737, 2x S23738, 2x S23739, 2x S23740, 2x S 23741, 2x S23742,
2x S23743, 2x S23744, 2x S23748, 2x S23750, 2x S23760, 2x S23775, 1x S23780, 2x S23790, 1x S23795,
2x S23805, 1x S23820, 1x S23825, 1x S23830, 1x S23835, 2x S23845, 2x S23860, 2x S23870, 1x S23880,
2x S23895, 3x S23900, 2x S23905, 1x S32130, 1x S32175, 1x S32200, 1x S32310, 1x S32375, 1x S93004,
2x S93030, 2x S93035, 2x S93050, 2x S93060, 1x S93063, 2x S93070, 2x S93075, 2x S93095, 2x S93100,
2x S93090, 1x Seite 39, 1x Seite 42, 1x Seite 13, 1x Seite 24

Index der Bildautoren
index of the photographers

U. Werner

1x S00106, 2x S07705, 2x S07717, 1x S13703, 3x S13735, 1x S13745, 1x S 13748, 1x S13755, 2x S13758, 1x S13762, 1x S13765, 2x S13760, 1x S13763, 2x S13770, 1x S23325, 1x S23330, 1x S23340, 2x S23375, 1x S23385, 3x S23455, 2x S23465, 2x S23500, 1x S23505, 2x S23510, 2x S23520, 1x S23580, 2x S23600, 3x S23635, 2x S23675, 1x S23705, 2x S23710, 1x S23715, 2x S23745, 2x Seite 34, 1x S23755, 2x S23758, 2x S23765, 1x S23780, 2x S23785, 3x S23810, 2x S23850, 2x S23863, 1x S23870, 1x S32105, 2x S32115, 1x S32120, 1x S32125, 2x S32140, 2x S32143, 3x S32145, 3x S32160, 1x S32185, 2x S32190, 3x S32195, 1x S32210, 1x S32235, 2x S32250, 2x S32260, 1x S32275, 1x S32280, 2x S32285, 2x S32320, 2x S32335, 3x S32345, 2x S32360, 2x S32365, 1x S32370, 2x S34000, 1x S34020, 2x S34035, 3x S34205, 1x S34210, 1x S34215, 1x S34220, 2x S34230, 2x S34240, 2x S34255, 2x S34275, 3x S78215, 1x S85305, 1x S85135, 3x S85325, 1x S85365, 2x S85370, 2x S85375, 2x S85385, 1x S85395, 3x S85405, 2x S93005, 2x S93010, 2x S39015, 2x S93020, 1x S93050, 1x S93075, 4x S99755, 1x Seite 65, 1x Seite 29, 1x Seite 30